Land and People.

The Russian Colonization of the Kazak Steppe

by

Gulnar Kendirbai

Contents

Introduction 1

§ 1 The Cossacks - the Pioneers of Russian Colonization 2

§ 2 The Peasant Colonization 13

§ 3 Colonial Land Policy 29

Conclusion 67

Bibliography 71

Introduction

The Russian colonization of the Kazak Steppe was characterized by the seizure of Kazak grazing lands by voluntary European peasants, mostly Russian Cossacks and peasants. This process began soon after the annexation of the Kazak Younger Horde in 1731[1], - long before the Kazak lands were officially declared state property given to the Kazaks for collective use in the Provisional Statute of 1868.[2]

The seizure of land led to a considerable decrease in pastures and the sharp reduction of livestock – the main wealth of the nomads – and caused a critical decline in the Kazaks' standard of living. In fact it had dramatic consequences for all aspects of Kazak traditional life, including those associated with any understanding of national identity.

However, this study will focus on the analysis of the character of colonial land legislation, namely how it became in practice arbitrary administrative rule, where any interference with established legal land relations not only proved illegal but also hampered the entire immigration campaign. Significantly, the above-mentioned Statute as well as the following Statutes of 1886 and 1891 did not contain any attempt at regulating immigration, thus favoring the spontaneous and haphazard character of Russian immigration. As a result, the process from the very beginning was controlled by local bureaucrats, who, finding themselves under strong pressure from the immigrants, took mainly provisional and occasional resettlement measures directed at the satisfaction of the immediate interests of the immigrants.

While focusing on only one, though the most important aspect of colonial legislation, one can get better

---

[1] The annexation of Kazak territories to Russia was completed in 1864.
[2] *Materialy po istorii politicheskogo stroia Kazakhstana*, vol. 1: *(So vremeni prisoedineniia Kazakhstana k Rossii do Velikoi Oktiabr'skoi Sotsialisticheskoi Revoliutsii)*, ed. M. G. Masevich, Alma-Ata, 1960, p. 337.

insights into the character and consequences of the *inorodtsy* policy conducted by tsarism in the empire's Asiatic outskirts. This, in turn, might contribute to a better understanding of the nature of the colonizer-colonized discourse and, hence, the reinvestigation of the degree of the latter's incorporation into imperial structures. This concern is of special relevance for the study of Russia's colonial history. From this perspective, the analysis of the diversity of colonial policies might provide historians with a good way of restoring the multiplicity of voices in the real imperial history of Russia.

§ 1 The Cossacks - the Pioneers of Russian Colonization

The immense and untamed spaces of the Kazak Steppe [3] that had neither settlements nor tilled land appeared before the eyes of the first newcomers as virgin lands untapped by any visible civilizing human activities. The nomadic lifestyle of their inhabitants, who continuously roamed the steppe, grazing their flocks, and had neither fixed defended frontiers nor a well-organized army capable of showing resistance, only strengthened the newcomers in their belief that it was "no one's land," thus facilitating to a certain degree their land seizure.

Since time immemorial, the Kazak Steppe had served as asylum for different kinds of fugitives: Russian peasants fleeing from serfdom, all sorts of freebooters, brigands and adventurers. They made their living by attacking and plundering both peaceful villages and trade caravans. The Ural'sk Cossacks since the late seventeenth century had been in a permanent state of war with Kazaks for the fertile valleys of the Ural River. The death penalty for civil crimes in the Russian empire was substituted by exile to Siberia and

---

[3] By the time under consideration Kazak tribes occupied territories historically associated with the Dasht-i-Kipchak steppes.

subsequent registration with the Siberian Cossacks. It should not be surprising then that the ethnic composition of the first Cossack groups was remarkably colorful: one could find among them captive Finns, Germans, and Poles, as well as Turks, representatives of Caucasus peoples, as well as Kalmyks, Tatars, Bashkirs, Mordva, Chuvash, and Albanians.

The Cossacks, a "free people", did not obey tsarist rule, and until the reign of Peter the Great had lived in accordance with their own understanding of law and order. With atamans and their assistants – (*esauls*), at their head elected by all members of their communities at general meetings – (the so-called *krug* or circle), they zealously maintained their freedom and independence. It was Peter the Great who, by using sometimes brutal measures, began to reduce their independence, forbidding the Cossacks to undertake independent military campaigns or accept fugitive peasants. Since 1700 their atamans were to be appointed by the tsarist government, and after 1738 by the tsar himself. In 1827, the crown prince was declared the ataman of all Cossack armies and preserved his title up to 1917. In the mid-eighteenth century, special Cossack guards under the personal command of the tsar were also formed.[4]

Gradually solidifying control over the Cossacks, tsarism at one time successfully used them in military campaigns, including those directed at broadening the empire's frontiers. Being "somewhere between subservience and freedom, between loyalty and independence"[5] the Cossacks participated in all wars conducted by tsarism: in the 1812 War, in the wars against Turkey, Sweden, Poland, Finland, and the Baltic peoples as well. They played an active role in the conquest of the Central Asian regions of Bukhara, Kokand, Khiva, and the Turkmen territories, as well as in the tsarist military operations in the Caucasus, Siberia, and the Crimea,

---

[4] Abdirov, M. J.: *Istoriia kazachestva Kazakhstana*, Almaty, 1994, pp. 97, 43, 26-27.
[5] Barrett, Thomas, M.: Crossing Boundaries: The Trading Frontiers of the Terek Cossacks, in: *Russia's Orient: Imperial Borderlands and Peoples, 1700-1917*, eds. Daniel R. Brower/ Edward Lazzerini, Bloomington/ Indianapolis, 1997, p. 244.

also including those carried out in the Far East and the Balkans. By the beginning of the twentieth century, the entire Cossack male population from the age of nineteen was liable to military service for nineteen years, including the provision of complete military equipment at their own expense; they were also to pay *zemstvo* taxes.[6]

By granting the Cossacks a special estate the tsarist government strove to secure their privileged position: they were released from all taxes and duties and granted property in land. Each Cossack from the age of seventeen could get thirty desiatinas[7] of land (the so-called *pai*) and hand it over to other Cossacks of his community for no more than one year, whereas the Russian peasants could officially be granted only ten to fifteen desiatinas. The land parcels of high-ranking servicemen were especially large: the land plots of the Siberian Cossack officers amounted to up to 400 desiatinas. The Ural'sk Cossacks held considerably larger land plots in comparison with other Cossacks: by the mid-nineteenth century, the plots of ordinary Cossacks reached something like 500 desiatinas, whereas those of their generals from 1,500 to 3,000 desiatinas. The lands remaining after the distribution among the members of Cossack communities were to form a reserve of the Ural'sk Cossack army and could be leased for various purposes for between 12 and 99 years.[8]

If by the eighteenth century the total strength of the Cossack armies was 85,000,[9] then by the mid-nineteenth century their figures had already increased to 160,000. By January 1st, 1881, the Cossack armies

---

[6] *Materialy po zemel'nomu voprosu v Aziatskoi Rossii*, vyp. 1. *Stepnoi krai*, ed. V. A. Tresviatskii, Petrograd, 1917, p. 87.

[7] measure of land = 2.7 acres; 0.405 hectares.

[8] O pozemel'nom ustroistve v kazach'ikh voiskakh, krome Ural'-skogo, in: *Sbornik zakonov i rasporiazhenii po pereselencheskomu delu i pozemel'nomu ustroistvu v guberniiakh i oblastiakh Aziatskoi Rossii (po 1 avgusta 1909 g.)*, SPb., 1909, pp. 451-454.

[9] Beskrovnyi, L. G.: *Russkaia armiia i flot v XIX veke*, M., 1973, p. 65.

included 51 mounted regiments, 290 mounted hundreds and squadrons, 37 infantry troops and 188 canons.[10]

The Cossack advance on the Kazak Steppe began with the building of military fortresses, outposts, redoubts, etc., and the occupation of lands around them, which were used for the organization of Cossack settlements - the so-called *slobodas* and *stanitsas*. First built as defensive outposts against nomadic forays, these fortresses were later also to be regarded as protecting Kazaks from the attacks of their troublesome neighbors. Fortresses and redoubts were connected with each other by mounted patrols, pickets and beacons, which formed a sort of line or military frontier that demarcated the territory occupied by their settlements. In this way, fourteen forts and eighteen outposts built by the Ural'sk Cossacks in the second half of the eighteenth century along the river Ural formed the so-called Ural'sk Military Line.

Estimated at 29,588 by the early nineteenth century and occupying a territory of about seven million desiatinas[11], the most prosperous group of the whole Cossack population of the empire, the Ural'sk Cossacks, had however long been in a search of new opportunities to enlarge their land properties, even provoking thereby the indignation of some colonial authorities. As G. Gens, the chairman of the Orenburg Frontier Commission (*Orenburgskaia Pogranichnaia Komissia*) established in 1782 by Catherine II, claimed to the Orenburg governor-general in 1830: "The land of the [Ural'sk] Cossacks, whose rights nobody questions, includes 7,000,000 desiatinas or about 3,000 geographical square miles, each mile of which falls to the share of twenty Cossacks, or, in other words, each male holds 500 desiatinas of land. Despite this, they continue to complain of their lack of land, though they themselves do not deal with farming at all as a rule." Another important source of the Ural'sk Cossacks' prosperity, according to the report, included fishing, from which they annually got at least 2,500,000 rubles

---

[10] Abdirov, M. J.: *Istoriia kazachestva Kazakhstana*, p. 29.
[11] Ibid., pp. 53, 55

while paying at the same time only six rubles of taxes per capita.[12]

Cossack farming was rapacious: they neither kept orders of plowing nor rotations nor even fertilization. This was characteristic and widely practiced by the Cossacks, especially at the earlier stage of their colonization, when plenty of "free" territories were relatively easily available, and the relatively quick exhaustion of soil inevitably forced them to capture new suitable lands. This trend of development had resulted in a dramatic decrease in fertile lands so that finally the Ural'sk Cossack administration forbade the Cossacks to leave their plots earlier than after three years. Each ordinary Cossack was allowed to plow no more than forty, while each high-ranking bureaucrat no more than eighty desiatinas, though in the end these measures proved unsuccessful.[13]

Though generally regarded as the common and indivisible property of their communities, the Ural'sk Cossacks preferred to practice farmstead (*khutor*) farming. However, usually they did not work their lands, as they came to widely practice the leasing of land to their former owners - the Kazaks. The latter, according to one account, paid for the wintering of their cattle at prices of fifteen kopecks per head of sheep, fifty kopecks per cow and eighty kopecks per camel in 1813.[14]

As early as the mid-eighteenth century, the Ural'sk Military Line was connected by a chain of new fortresses with the cities of Orenburg and Orsk, which had also been built up earlier as military forts. Later on, the construction of a line of new fortresses was continued up to the Ural Mountains and the Russian Tobol'sk province. A new joint line that extended from the upper reaches of the Ural to the Ural Mountains for

---

[12] *Kazaksko-russkie otnosheniia v XVIII – XIX vekakh (1771-1867 gody), (Sbornik dokumentov i materialov)*, eds. M. O. Jangalin/F. N. Kireev/V. F. Shakhmatov, Alma-Ata, 1964, pp. 238-239.

[13] Borodin, A.: *Ural'skoe kazach'e voisko. Statisticheskoe obozrenie v 2-kh tomakh*, vol. 1, Ural'sk, 1891, pp. 477-478.

[14] Abdirov, M. J.: *Istoriia kazachestva Kazachstana*, p. 54.

1,780 versts [15] and included altogether forty fortresses, outposts and *stanitsas* was named the *Orenburgskaia Voenno-Pogranichnaia liniia* (the Orenburg Military-Frontier Line). From the fortresses of Iletskaia krepost' to Zverinigolovskaia it demarcated the territory of the Orenburg Cossacks - the second largest Cossack population on the steppe and the third largest in the empire, after the Don and the Kuban Cossacks, who by the beginning of the nineteenth century numbered more than 55,000. By 1917 their population had already grown to 533,000 people, and the amount of their lands to more than 7,4 million desiatinas.[16]

The construction of the two earlier fortresses of Iamyshevskaia and Omskaia of the Irtysh Military Line on the right-bank of the Siberian River Irtysh began in 1715 -1716. In the following years, the line was extended to the east-southern interior of the steppe by the construction of new forts. From the opposite side it was connected with the Orenburg Military-Frontier Line by nine fortresses and fifty-three redoubts of the New Ishim Line. By the mid-eighteenth century, the Irtysh, the New Ishim and the Kolyvano-Kuznetsk Military Lines[17] formed a 2,991 verst long Siberian Military Line - a military cordon of altogether eighteen fortresses, thirteen outposts, thirty-one redoubts, twenty-three stations, and thirty-five beacons that outlined the territory of the third largest Cossack group - the Siberian Cossacks, blocking at the same time the path of the Kazaks of the Middle Horde to their summer pastures located between the Rivers Irtysh, Esil and Tobol, as well as to the right-bank of the River Irtysh.[18] The Siberian Cossacks were also granted the exclusive right to fish in the River Irtysh and Lake Zaisan.

In accordance with the 1846 Regulations that had been prompted by the rebellions headed by Kenesary Qasymov, each Siberian Cossacks field officer was

---

[15] measure of land = 3,500 feet or 1,06 km.
[16] Ibid., p. 87.
[17] The building of the Kolyvano-Kuznetsk Military Line began in 1745. Its nine forts and 53 redoubts connected the Irtysh riverside with the region of Altai.
[18] Ibid., p. 96.

permitted to enlarge his private lands to 400 desiatinas, whereas each head-officer was allowed 200 desiaitinas, whereby each ordinary Cossack could also be additionally granted twenty-five desiatinas. In case of land shortage the land could be withdrawn from Kazak pastures. Moreover, thirteen new Cossack settlements *(stanitsas)* comprised of forcibly resettled Cossacks and 5,000 peasant settlers previously registered with the Cossack estate were also to be established, whereas the non-Cossack population living in the territory of the army was to leave it within two years.[19]

As always, though owning big land plots, the Siberian Cossacks did not till them, preferring to lease them to Kazaks, who were exploited as gratuitous workers. As reported by Shcherbina, the land of the Siberian Cossacks was divided into the plots of soldiers *(iurtovye nadely)* and their officers, as well as those who had been granted land instead of pensions in the territory of the ten-verst strip. The third group constituted the army's reserve lands. All these lands had been held on lease by the local Kazaks, ninety-three percent of whom had been living on the plots of the first two types. Since the army's reserve constituted mainly infertile lands, they had been put out to lease to poor Kazaks who paid a considerably lower rent. While ordinary Cossacks leased their plots in indivisible portions, their officers preferred to lease their lands in several separate plots, so that they could be used for farmstead farming, haymaking and pastures as well. Thereby the officers often used the *kulaks* (the prosperous peasants) as their mediators in lease operations with Kazaks.[20] In 1901 the number of Kazaks of only three districts of the Semipalatinsk

---

[19] Katanaev, G. E.: *Kirgizskii vopros v Sibirskom kazach'em voiske*, Omsk, 1904, p. 10.

[20] *Materialy po kirgizskomu zemlepol'zovaniiu, sobrannye i razrabotannye ekspeditsiei po issledovaniiu stepnykh oblastei*, vol. IV: *Semipalatinskaia oblast', Pavlodarskii uezd*, Voronezh, 1903, pp. 42-43.

province who leased the land of the Siberian Cossacks under hard conditions was said to be 12,000 people.[21]

All the same, the Siberian Cossack commanders constantly sought to extend their territories. From governor-generals to the Ministry of War they kept complaining to various authorities about the lack of fertile lands causing "great discomfort and need" as well as Kazak livestock "surrounding their settlements from all sides" and damaging their fields and leading to strip farming, as did those of them inhabiting the Altai region. Since "the Cossacks as a military people were not used to yielding without struggle," their petitions were in the end complied with, despite the protests of the Altai Kazaks and even some local bureaucrats.[22]

It was estimated that by 1917, 172,000 Siberian Cossacks held altogether 4,762,418 desiatinas of land that averaged twenty-seven to forty-three desiatinas per male capita, whereas the land properties of the Ural'sk Cossacks comprised 6,465,402 desiatinas, or on average 79 desiatinas per male capita.[23]

The fourth largest Cossack group in Central Asia was the Semirech'e Cossacks, whose first fortification Vernyi (modern Almaty) was built in 1854 in the southern Kazak province of Semirech'e (Jetisu, literally Seven Rivers) that in the second half of the nineteenth century formed a part of the governor-generalship of Turkestan along with the Syr Darya province. The region, due to its exceptionally favorable climatic conditions – a warm climate and fertile soil along with an abundance of water and forests – quickly became a popular place of settlement in the eyes of both the Cossack and peasant settlers. The latter, coming from less climatically favorable places in European Russia, were

---

[21] Suleimenov, B. S.: Rabochee i agrarnoe dvizhenie v Kazakhstane v 1905-1907 godakh, in: *Voprosy istorii Kazakhstana XIX - nachala XX veka*, ed. T. E. Eleuov, Alma-Ata, 1961, p. 7.

[22] Shmurlo, E.: Russkie i kirgizy v doline Verkhnei Bukhtarmy, in: *Tsarskaia kononizatsiia v Kazakhstane (Po materialam russkoi periodicheskoi pechati XIX veka)*, ed. F. M. Orazaev, Almaty, 1995, pp. 68-69, 78.

[23] *Materialy po zemel'nomu voprosu v Aziatskoi Rossii*, vyp. 1. *Stepnoi krai*, pp. 86-87.

often registered with local Cossack communities. Some 300 former peasant families, who had come in 1859 from the Tomsk province, formed a new Cossack force which was called the Zailiiskii battalion. They were released from military service, including all taxes and duties for two years. In addition, each family was granted fifty-five rubles and thirty desiatinas of land.
Consequently, by the 1867 decree of Alexander II, an independent Semirech'e Cossack Army was formed on the basis of the ninth and the tenth regiments of the Siberian Cossack Army. Later, it actively participated in the conquests of the khanate of Khiva in 1873 and Kokand in 1875-1876, as well as the cities and forts of southern Kazakstan: Merke, Pishpek, Tashkent, Shymkent, including also the military campaigns in the western part of China. For the successful implementation of these military operations the plots of ordinary Cossacks were enlarged to fifty desiatinas: In some places the Semirech'e Cossacks held up to 300 desiatinas of land per male capita instead of the legal thirty.[24]
According to the data of the Governor-General of Turkestan, Konstantin von Kaufman, by the mid-nineteenth century, forty percent of the Semirech'e peasants leased land from the Cossacks. Of a total of 820,650 desiatinas in the possession of the sedentary population of Turkestan, 629,896 desiatinas fell to the share of the Semirech'e Cossacks; the figures for the peasant and indigenous sedentary populations were 60,000 and 170,000 desiatinas respectively.[25]
Under the rule of the Governor-General of Semirech'e, G. A. Kolpakovskii, the plots of the local Cossacks were limited to twenty desiatinas per male capita in the 1880s. Apart from this, some newly established Cos-

---

[24] Abdirov, M. J.: *Istoriia kazachestva Kazachstana*, pp. 121-125.
[25] Ivanov, A: Russkaia kolonizatsiia v Turkestanskom krae, in: *Tsarskaia kolonizatsiia v Kazakhstane*, pp. 13-14, 18. Count K. K. Pahlen during his revision in Turkestan in 1908 observed the families of retired officers of the Russian Army amusing themselves until morning by singing, drinking and playing cards at the casinos of the city of Shymkent; see: Pahlen, Constantin Graf von der Pahlen: *Im Auftrag des Zaren in Turkestan 1908-1909*, Stuttgart, 1969 [Bibliothek klassischer Reiseberichte], p. 285.

sack settlements, despite the protests of their inhabitants, were simply declared peasant, and all inhabitants registered with peasant status.[26] Consequently, the local administration managed to cut off altogether 215,000 desiatinas in favor of land-hungry Kazaks and Russian peasants, 100,000 desiatinas of which were given back to the Kazaks of the Lepsy district.[27] Later, however, the Ministry of War found this decision incorrect and suggested that 116,000 desiatinas from the total of 130,000 desiatinas of the army's reserve could be left to the voluntarily immigrated peasants, providing they registered with the Cossack estate.[28] Nevertheless, the Semirech'e Cossack administration had in the end found other ways to restore its decreased land properties, so that by 1917 more than 45,000 Semirech'e Cossacks were reported to be in the possession of altogether 681,000 desiatinas of land.[29]

With time, all Cossack groups settled on the steppe assumed a punitive role, actively participating in the suppression of native uprisings – most notably, the rebellions led in 1843 by Kenesary Qasymov and by Eset Kotibarov in 1855. They were also effectively employed in the suppression of the 1916 revolt that had embraced the major Kazak regions.[30] The 1916 revolt in Central Asia was provoked by the tsarist Ukaz of 25 June, 1916, which attempted to mobilize 400,000 Central Asian men between the ages of 19 and 43 for the World War I war effort.[31] In peacetime, however, the Cossacks often initiated punitive expeditions to Kazak *auyls* (nomadic camps) on their own, so as to

---

[26] Ivanov, A: Russkaia kolonizatsiia v Turkestanskom krae, p. 333.
[27] *Zhurnal Soveshchaniia o poriadke kolonizatsii Semirechenskoi oblasti*, Vernyi, 1908, pp. 41, 44
[28] Ibid., p. 41.
[29] Abdirov, M. J.: *Istoriia kazachestva Kazachstana*, p. 132
[30] Ibid., pp. 57, 105-107.
[31] Before World War I Central Asian men were not called to military service in the Russian army. Altogether 1.5 million Muslim soldiers served in the regular Russian army at this period. They were recruited from the Caucasus and other sedentary Muslim areas, and they were made up mainly of Tatars; see: S. M. Iskhakov: Pervaia mirovaia voina glazami rossiiskikh musul'man (unpublished manuskript).

capture cattle and other things, but also women, of whom, especially at the beginning of Cossack colonization, their communities had often been in want.

In his letter sent to Catherine II in 1790, Syrym Batyr, the leader of the Kazak uprising in the Younger Horde, complained about the systematic killings and plundering raids practiced by ataman Danila Dashkov and his Ural'sk Cossacks. During one such attack his 1,500 Cossacks killed 150 Kazaks and took another 57 Kazaks as prisoners; they destroyed and seized the belongings of 225 yurts along with numerous horses, camels, cows and sheep as well. [32]

Given the mass character of these attacks, which as a rule were accompanied by murder and the seizure of large numbers of Kazak livestock, the tragic conesquences of these expeditions finally drew the attention of the Asiatic Department of the Ministry of Foreign Affairs. By a special order of September 21, 1814, it prohibited the Ural'sk Cossack leaders from dispatching expeditions to the steppe. [33]

In a similar pattern, the Orenburg local authorities in 1858 reported regular forays carried out by the Orenburg Cossacks on their own initiative aimed at cattle seizure. As it was, a year before they "punished" several dozen *auyls* for their alleged participation in the riots in the Younger Horde headed by Janqoja Nurmukhamedov by taking away 21, 400 of their livestock. [34]

Though incorporated into the basic administrative units of the Steppe and Turkestani governor-generalships - *gubernii* (provinces), in accordance with the 1868 and 1886 Statutes, the Cossack administrations, however, were granted wide powers making them equal to the provincial administrations (*gubernskoe pravlenie*) of the Russian European provinces, namely they were to perform the duties of a fiscal chamber, local civil and criminal courts, as well as the regional administrations of state properties. The governor-generals of the

---

[32] *Materialy po istorii Kazakhskoi SSR (1785-1828 gg.)*, vol. IV, M/L, 1940, p. 137.

[33] Abdirov, M. J.: *Istoriia kazachestva Kazakhstana*, p. 55.

[34] *Kazakhsko-russkie otnosheniia v XVIII – XIX vekakh (1771-1867 gody)*, pp. 429, 443.

provinces of Ural'sk, Semipalatinsk and Akmolinsk were entitled to perform the duties of the atamans of the Ural'sk and Siberian Cossack Armies. In addition, the 1868 Statute preserved their local courts (*stanichnye sudy*) for small crimes. More serious crimes were to be placed under the general imperial courts represented by the military judicial commissions (*voenno-sudebnye komissii*), as well as the uezd and oblast administrations. [35] However, in contrast to the latter that in accordance with the 1864 Statutes of Alexander II were to be elected by communities themselves, the Cossack uezd courts were to be appointed by the government and approved by the Minister of Justice. Thus, Cossack judicial issues, including also their police and military, were directly subordinated to the central ministries, while the Steppe and Turkestani governor-generals were to perform only the role of mediators between them and the central authorities. [36]

All in all, by the beginning of the twentieth century new Cossack domains occupied a territory of about twenty million desiatinas representing separate economic, administrative-territorial and social units within the established colonial administrative units. Fenced in by a cordon of military fortresses and redoubts that formed a kind of military frontier, they cut off the territory occupied by Cossack settlements from the rest of the steppe inhabited by both Kazaks and Russian peasants.

§ 2 The Peasant Colonization

The immigration of settlers from European Russia, predominantly the Russian peasants - the second largest group of migrants - began mainly in the second half of the nineteenth century. The Great Reforms of the 1860s

---

[35] *Materialy po istorii politicheskogo stroia Kazakhstana*, pp. 323-325, 329-330.
[36] Borodin, A.: *Ural'skoe kazach'e voisko. Statisticheskoe obozrenie v 2-kh tomakh*, vol. 1, pp. 253-255, 12-13.

that envisaged the emancipation of the Russian peasantry contributed to the growth of their general mobility.

It has been estimated that from 1861 to 1885 more than 300,000 peasants crossed the River Ural.[37] Coming mainly from the Russian and Ukrainian provinces of Poltava, Samara, Astrakhan, Voronezh, Chernigov, Penza, Saratov, and Kazan, the first newcomers, as mentioned above, were often registered with Cossack status or with the citizens of local towns and in that way released from taxes and duties at their previous places of residence.

The settlers, comprising peasants and workers fleeing from unbearable conditions of work and life, as well as religious refugees, exiles, including fugitives from jails and penal servitude along with various kinds of criminals, as a rule did not hesitate to capture the most suitable lands, settling themselves down at once in new places. This being the case, the settlers who appeared in 1866 in the southern part of the Altai region began mowing grass on the Kazak meadows, plowing fields and building houses and bee-gardens, without first asking for any permission or directions. But more than this, as one of the inhabitants of the region, the Kazak Qaratai tribe, complained in their appeal, the newcomers took their winter pastures, burnt their winter camp buildings, and even the grass on the steppe in order to chase them out of their pastures. In the eyes of some colonial authorities, the organization in such a manner of the first Russian villages of Medvedka and Talovka was to serve as proof of the "sharpness, persistence, and talent of the Russian man as a colonizer, and at the same time the weakness of *inorodtsy*, who were unable to oppose the penetration by a strange element." The chief of the Ust' Kamenogorsk district colonial administration of the Altai region also reported the unbidden guests who had set up their arable fields on the local Kazak pastures, plowed all nomadic roads

---

[37] Bekmakhanova, N. E.: *Mnogonatsional'noe naselenie Kazakhstana i Kirgizii v epokhu kapitalizma (60-e gody XIX v. – 1917 g.)*, M., 1986, p. 93.

in order to prevent the native inhabitants from using their summer pastures, and immediately complained about the inevitable damage caused to their fields by crossing Kazak flocks.[38]

Of the four steppe provinces it was Akmolinsk province that became especially popular with the Russian peasants: the first voluntaries from the Russian provinces of Tobol'sk and Perm appeared there as early as in 1866. They took on lease pieces of Kazak land and established on their own initiative a settlement of fifty houses. In the 1870s, by the order of the Chief Administration of Western Siberia (*Glavnoe Upravlenie Zapadnoi Sibiri*), 819 males were allotted altogether eighteen plots by granting each of them thirty desiatinas of land.[39] By 1881, in the wake of the implementation of the "Provisional Regulations on the Immigration of Rural Inhabitants to the Kirgiz [Kazak][40] Steppe" (*Vremennye pravila po pereseleniiu v kirgizskie stepi sel'skikh obyvatelei*), the total number of immigrants resettled in the province reached 3,650.[41] Within six years the Russian population of the province had already grown to 25,000. During only one year, when the Gossovet by its 1889 law officially permitted the immigration to the "free state lands," 353 families (2,238 people), mainly from the Russian provinces of Samara and Saratov, settled down in Akmolinsk province. Despite the provisional suspension of immigration in 1891 ordered at the request of the Steppe governor-general, it continued to grow and became especially acute after the bad harvests of the 1890s in Russia. The total figure for peasant population immigrated to the province between 1860 and 1896 was calculated to amount to 98,517.[42]

---

[38] Shmurlo, E.: Russkie i kirgizy v doline Verkhnei Bukhtarmy, pp. 29, 50.
[39] Bekmakhanova, N. E.: *Mnogonatsional'noe naselenie Kazakhstana i Kirgizii v epokhu kapitalizma*, p. 59
[40] "Kirgiz" was a term employed throughout imperial Russian period to designate Kazaks.
[41] Shonauly, Teljan.: *Jer tagdury – el tagdyry*, Almaty, 1995, p. 137.
[42] Bekmakhanova, N. E.: *Mnogonatsional'noe naselenie Kazakhstana i Kirgizii v epokhu kapitalizma*, p. 93.

In like manner, even the official closure of Ural'sk province did not prevent 5,480 peasant families from resettling there in the period 1867-1897. By 1915, their number had already increased to 82,900 and comprised 9.7 percent of the province's total population. The first four peasant settlements in Semipalatinsk province were set up in 1891; they were allotted something like 33,000 desiatinas of land. By 1897, however, 4,570 peasants were already registered as resettled in the province.

The most actively immigrated places of Turgai province included the Kustanai and Aktiubinsk uezds (districts): the colonization of the first of them began in 1880. If by 1889 the settlers comprised a figure of 20,000,[43] by the beginning of 1897 they already amounted to 30,000.[44]

The 1881 Provisional Regulations were followed by the "Provisional Regulations on the Creation of Settlement and Reserve Plots" of June 13, 1893, as well as the subsequent laws of the Gossovet concerning the creation of settlement plots in the steppe provinces: first promulgated on April, 12, 1904, for Ural'sk province, later on February 14, 1905, for the Semirech'e province and, finally, on January 2, 1906, for the provinces of Akmolinsk, Semipalatinsk and Turgai.[45]

Significantly, from a total of 1,117,389 Russian peasants settled by 1905 in all steppe provinces, only 117,316 or twelve percent returned to Russia.[46]

Due to its extraordinarily fertile soil, Semirech'e and another Kazak province of Turkestan, Syr Darya, quickly became the most popular places for colonization. Count Pahlen believed that the Land of Seven Rivers was not the province's true name, but a land "flowing with milk and honey," a kind of Central Asian Eldorado that "teemed with pheasants, wild hares,

---

[43] Shonauly, T.: *Jer tagdury – el tagdyry*, p. 138.

[44] Bekmakhanova, N. E.: *Mnogonatsional'noe naselenie Kazakhstana i Kirgizii v epokhu kapitalizma*, pp. 115-116.

[45] RGIA (The Russian State Historical Archive), f. 1276, op. 4, d. 468, l. 218.

[46] Shonauly. T.: *Jer tagdury – el tagdyry*, p. 141.

rabbits and other birds," where man's slightest efforts were rewarded a thousand-fold by nature.[47]

Though the province had also been officially closed for immigration since 1861, a stream of unauthorized immigrants had been permanently growing, despite desperate attempts by the local bureaucrats to suspend it: in 1896 their population was placed at 40,000. It was, however, not until the beginning of the twentieth century that the province suffered its biggest influx of unauthorized immigrants: If in 1902 the newcomers accounted for 16,000, by 1905 their figure had already reached 23,000. In 1907 and 1908 their population was estimated at 24,700 and 30,000 respectively,[48] in spite of an official immigration ban in 1907.[49]

High immigration was also characteristic of the Syr Darya province, the colonization of which began in 1876 with the resettlement of more than 2,000 Russian peasants. By 1891, they were reported to already number 16,400.[50]

A. A. Kaufman, a member of the local resettlement office, singled out the three main groups of Turkestani settlers. The first was represented by former soldiers of the Russian Army, who after retiring from their service, could remain in Turkestan, be granted land and a hundred rubles, as well as be released from land taxes.[51] Two other groups included peasants coming from various Siberian and steppe regions, and those emigrating directly from Russia's European provinces. As a rule, the former soldiers settled in Tashkent and put their land out to lease. In turn, peasants from

---

[47] Pahlen, C.: *Im Auftrag des Zaren in Turkestan 1908-1909*, p. 263.

[48] Pahlen, K. K.: *Pereselencheskoe delo v Turkestane. Otchet po revizii Turkestanskogo kraia, proizvedennyi po vysochaishemu poveleniiu senatorom gofmeisterom grafom K. K. Palenom*, SPb., 1910, p. 17.

[49] RGIA, f. 1276, op. 4, d. 468, l. 212.

[50] Bekmakhanova, N. E.: *Mnogonatsional'noe naselenie Kazakhstana i Kirgizii v epokhu kapitalizma*, p. 110.

[51] Pahlen during his revision in 1908 observed the families of retired officers of the Russian Army, amusing themselves until morning by singing, drinking and playing cards at the casinos of the city of Shymkent. *Im Auftrag des Zaren in Turkestan 1908-1909*, p. 285.

Siberia and the steppe regions represented the largest and at the same time most unstable group of settlers. After arriving in Turkestan, they would change their places of residence up to ten times, if they learned of better places to settle. By contrast, the peasants coming from Russia's European provinces usually followed their own relatives who had settled earlier. The old settlers as a rule provided them in advance with necessary information concerning climatic conditions, organizational questions, etc.[52] As reported by Pahlen, forty percent of all settlers in Turkestan were Siberian immigrants[53]: in June of 1900 alone one million of them came by the Western Siberian Railway,[54] a section of the Trans-Siberian Railway completed in 1897 that connected Central Siberia with Orenburg via Omsk, Petropavlovsk, Cheliabinsk and Samara. It was unified with the Central Siberian line into the "Siberian Railway" in 1900.[55]

It was, however, not until the implementation of the Stolypin agrarian reform (1906-1911) that the immigration of Russian peasants to the idle lands of Kazakstan and Siberia assumed an acute and dramatic character. By 1916, the number of peasants that had migrated to the six Kazak provinces (Turgai, Akmolinsk, Ural'sk, Semipalatinsk, Syr Darya and Semirech'e) made up 1,367,000 people or 23.1 percent of the whole population.[56] In the period of eight years from 1906 to 1913 in Semirech'e province alone about three million desiatinas of land were withdrawn.[57] The reform aimed at the destruction of the Russian rural community (*obshchina*) and the creation of strong and independent family farms *(khutory)*. It allowed poor peasants to sell

---

[52] Kaufman, A. A.: *K voprosu o russkoi kolonizatsii Turkestanskogo kraia. Otchet chlena Uchenogo komiteta Kaufmana po komandirovke letom 1903 g.*, SPb., 1903, pp. 8-9.
[53] Pahlen, K. K.: *Pereselencheskoe delo v Turkestane*, p. 14.
[54] Shonauly, T.: *Jer tagdury – el tagdyry*, p. 121.
[55] Marks, Steven, G.: *Road to Power. The Trans-Siberian Railroad and the Colonization of Asian Russia 1850-1917*, Ithaca/New York: Cornell University Press, 1991, p. 191.
[56] Qoigeldiev, Mambet: *Alash qozgalysy*, Almaty, 1995, p. 51.
[57] Abdirov, M. J.: *Istoriia kazachestva Kazachstana*, p. 129.

and purchase their lands, thereby favoring the concentration of lands in the hands of the *kulaks* and stimulating at the same time the emigration of the poor to other provinces of the Russian empire, in particular to Central Asia and Siberia. Emigrating peasants were released from all taxes (for fifteen years) and all duties (for twenty-five years); in addition, each family received a loan of a hundred rubles and each man was granted thirty desiatinas of land.

Contemporary accounts of the main reasons for immigration, indeed, did not much differ in character from those made a half century earlier, pointing to the continuing impoverishment among a large portion of the Russian peasantry as well as periodical bad harvests:

> It is, indeed, like a huge river of people rushing from native Russia to the Trans-Ural Region. Like the birds of passage irrepressibly flying in huge flocks with the beginning of cold weather to the warm overseas lands, so the peasants after becoming starved and impoverished in their own home land, have been in search of freedom and prosperity in an unknown land.[58]

Whereas peasants immigrating without authorization simply captured the first land they liked, others followed the so-called *khodoki* (scouts) - the representatives of their own *obshchina* sent in advance for the investigation of fertile lands and the conclusion of lease agreements. Stolypin's government promoted the institution of *khodoki* by introducing special permissions allowing *khodoki* to emigrate that were to be given to them by local authorities. After choosing their lands, they were to hand their permission papers over to local resettlement bureaucrats who, for their part, were to preserve their lands for their families for two years. Accordingly, the *khodoki* were to be released from military service for three years as well as from all taxes and duties for the first five years. In the following five years, they were to pay only a half of taxes. Each

---

[58] Glebov, A.: *Chto mogut dat' pereseleniia krest'ianstvu*, SPb., 1907, p. 2.

family was also to be granted 165 rubles and a ticket. Their land plots left at home could be passed on to their former communities or the latter's members for a certain payment. Families immigrated without permission were deprived of all these privileges and could be resettled only after the resettlement of the *khodoki*'s families.[59]

The following data of the Resettlement Office show the dynamics of immigration beyond the Urals between 1896 to1909[60]:

Years	Immigration	Emigration
1896	178,000	21,000 or 13%
1897	168,000	18,000 or 12%
1898	148,000	18,000 or 12%
1899	170,000	21,000 or 12%
1900	166,000	42,000 or 25%
1901	89,000	33,000 or 37%
1902	82,000	26,000 or 31,5%
1903	94,000	21,000 or 22%
1904	40,000	10,000 or 25%
1905	38,000	8,000 or 21%
1906	139,000	13,000 or 9,7%
1907	427,000	27,000 or 6,3 %
1908	665,000	45,000 or 6,7 %
1909	619,000	82,000 or 13%

It was, indeed, a long and harsh way to reach the remote and unknown Asian provinces: It was said that peasants emigrating to Turkestan from Russia's European provinces needed at least two to three years to get to the Promised Land. Exhausted by their tough journey and, as a rule, being upon arrival in the possession of neither cattle nor capital and agricultural implements, and not even seed, they were granted by local

---

[59] *Pereselenie v Stepnoi krai v 1906 g. (oblasti Akmolinskaia i Semipalatinskaia)*, vyp. 27, pp. 56-58.

[60] Quoted in: *Bökeikhan, Alikhan: Tandamaly/Izbrannoe: Nauchnye issledovaniia, trudy na kazakhskom i russkom iazykakh: monografii, stat'i, rechi, doklady, stikhi, perevody, pis'ma*, ed. S. Aqqulyuly, Almaty, 1995, p. 373.

resettlement offices only twenty-five rubles.[61] By contrast, the experienced Siberian settlers were arriving in Turkestan in wagons well adapted to long journeys and loaded with home necessities and baggage.[62]

The Akmolinsk Resettlement Office reported in 1907 that peasants came to the province from all directions: by the Siberian railway, by carts from Turgai province to the districts of Kokchetav and Atbasar, as well as on the River Irtysh by steamers from the cities of Omsk to Pavlodar, and then by carts to the Akmolinsk district. Those who went by cart received only insignificant financial support from local resettlement offices, whereas those who used the railway could be granted meals at public expense at the railway stations of Petropavlovsk, Omsk and Issykkul. The last got some milk, bread and *shchi* (cabbage soup) from the middle of March to the middle of October, and from October to the end of the year - bread and milk for their children under five years of age, as well as tea with sugar and bread for older children and themselves. In all, in the year under consideration 126,469 immigrants received altogether 202,000 portions at a value of 6.2 kopecks each.[63]

Not surprisingly, the beginning of a new life proved for many settlers a test of their ability to survive under often very hard conditions. As accounted by the Steppe governor-general: "Settlers live during their first years in their narrow, damp and cold houses and suffer terrible hardships.... Because of the absence of building materials, they dig dugouts resembling holes rather than places of human habitation. Various illnesses developed under these conditions – including malnutrition – and claimed many of their lives."[64] The statistician of the Syr Darya Resettlement Office, I. Geier, after visiting several settlers' houses in 1891,

---

[61] Ivanov, A.: Russkaia kolonizatsiia v Turkestanskom krae, p. 19.

[62] Pahlen, K. K.: *Pereselencheskoe delo v Turkestane*, p. 14.

[63] *Otchetnye dannye po Akmolinskomu pereselencheskomu raionu za 1907 g. na osnove otcheta byvshego zav. pereselencheskim delom v Akmolinskom raione Reznichenko*, vyp. 49, SPb., 1908, pp. 29-29.

[64] Quoted in: Bekmakhanov, E. B.: *Prisoedinenie Kazakhstana k Rossii*, M., 1957, p. 281.

wrote about his impressions in one of them: "One's heart bleeds on entering his [the settler's] house built only last autumn. The walls ooze with dampness, the ceiling is leaking; only a few houses have tables, others instead use heaps of stones covered by clean tablecloths. Despite all this, everybody is so bright and cheerful, and so full of belief in their future happiness!" Another contemporary witnessed the building of houses in the Russian settlement of Poltavskii:

> Building materials used by them were limited to small brushwood, soil and clay...They all say that if they had been given wood, they would have built the whole city of Petropavlovsk. All women trample wettened earth with their bare feet to make clay for the walls of the dugouts. Small children of seven to ten years of age carry water in pails from the lakes. If the pail is very heavy, a child ties it to a long pole, so that the longer end of the pole rests against the ground, while its shorter end he carries himself.

Accordingly, high mortality caused by hunger, food shortages as well as illnesses became an everyday occurrence among them: one account estimated that during only one year about a third of the 363 families of the settlement (*selo*) Mikhailovskoe had died from various illnesses in 1895.[65]

Yet, those of them who managed to overcome the difficulties of the initial period appear to have successfully established themselves. The statistical expedition under the leadership of K. V. Kuznetsov reported that an average peasant family of the Akmolinsk and Omsk uezds of Akmolinsk province populated by a total of 9,890 settlers was in the possession of eight head of cows and horses and seven head of small cattle in 1911.[66]

Similarly, settlers in Akmolinsk province, observed by V. A. Tresviatskii in 1916, showed a remarkable in-

---

[65] Quoted in: Suleimenov, B. A.: *Revoliutsionnoe dvizhenie v Kazakhstane v 1905-1907 godakh*, Alma-Ata, 1977, pp. 36-38.
[66] TsGA RK (The Central State Archive of the Republic of Kazakstan), op. 1, d. 18, l. 9.

crease in their cattle and agricultural implements: an average family, according to his categorization, possessed eight head of large livestock, which included cattle and horses, and seven head of smaller livestock, which may have included sheep, goats and pigs. According to 1916 provincial data, sixty percent of the families were completely equipped with agricultural implements. The total profit for an average peasant household was made up of fifty-seven to sixty percent from land, eighteen to twenty percent from cattle, fifteen percent from handicrafts and ten percent from other kinds of profits, though, generally, handicrafts among the peasants of the steppe provinces were poorly developed. The cash profit of an average household was estimated at 4,714 rubles in 1911, whereas its expenses made up 2,574 rubles.[67]

Yet the settlers of the Kazak provinces of Turkestan had enjoyed the fastest increase in peasant prosperity. A. A. Kaufman, who visited the Syr Darya province in 1903, was quoted as saying: "As for my private impresssions, I had to frankly admit that during my frequent journeys to various colonial regions I have nowhere met such collective prosperity and full satisfaction with life as in the settlements of the Syr Darya province." There were no poor among those settlers emigrated mostly from the northern regions of the Turgai province: as a rule, they were very well equipped with agricultural implements, working cattle etc., and gathered rich harvests. This being the case, 120 families from the settlement of Georgievskii were in possession of 1,000 draught oxen and two herds of horses. Each family owned two to three horses and a plough, while the richest of them had, in addition, 2,000 to 3,000 rubles. As Kaufman remarked, an average peasant plot was four to five times larger than those owned by Turkestan's indigenous populations.[68]

---

[67] *Materialy po zemel'nomu voprosu v Aziatskoi Rossii*, vyp. 1, *Stepnoi krai*, p. 102.
[68] Kaufman, A. A.: *K voprosu o russkoi kolonizatsii Turkestanskogo kraia*, pp. 112-113.

All the same, provided the vast spaces of the untouched Kazak Steppe, the peasants' prosperity, however, was basically enabled by the unlimited and in most cases unproblematic lease of Kazak lands.[69] Applied to the Central Asian soil, the rapacious and primitive kind of farming that the settlers had brought from their native provinces had finally taken its toll. As accounted by Pahlen, in the course of thirty years, the settlers farmed out the soil of the exceptionally fertile Lepsy region of Semirech'e to such a degree that instead of wheat only weed would grow. Beyond this, as both Pahlen and A. A. Kaufman indicated, despite the local favorable climatic conditions, the settlers did not cultivate gardens, grow vineyards or tobacco, preferring wheat to all other sorts of grain.[70] Similarly, another observer claimed that the Russian settlers in Turkestan did not learn how to use the local climatic conditions from the local sedentary population: they continued to grow wheat and potatoes as well as mow grass, without thinking of profit and making any efforts to produce products for sale in order to have additional money for the satisfaction of their other needs, adding that: "All of them look like orphans seeking guardianship without which they think they could not manage; nowhere did we hear that they felt happy or were satisfied with their new life; we neither saw joyful faces nor heard happy speeches. None of them has put down roots in their new land. All are ready to get under way and move to new places, to 'lands flowing with milk and honey'." The absence of private initiative and solidarity, the author believed, made the colonists, "despite their satiety, the most pitiful people."[71]

Yet, instead, the settlers appeared to capitalize on some distinctive patterns of behavior and mentality that, as we have seen, had already distinguished their prede-

---

[69] *Zhurnal Soveshchaniia o poriadke kolonizatsii Semirechenskoi oblasti*, p. 32.

[70] Pahlen, K. K.: *Pereselencheskoe delo v Turkestane*, pp. 148-150; Kaufman, A. A.: *K voprosu o russkoi kolonizatsii Turkestanskogo kraia*, p. 44.

[71] Ivanov, A.: Russkaia kolonizatsiia v Turkestanskom krae, pp. 17-18.

cessors. Count Pahlen cited as a typical example the peasant settlers who, after ignoring all assertions of the local bureaucrats about the absence of free lands, illegally took pieces of Kazak land on lease, and after establishing themselves near the old peasant settlements, invited their relatives and acquaintances, finally demanding local administrations to turn the leased pieces of land into their private property. Then, after privatizing these lands, they sold them to new settlers and moved on to other places with better soil where they applied the same trick. Plots could be sold even without previous privatization, as was the case with the Semirech'e peasants. The latter, instead of paying from twenty-seven to 170 rubles per desiatina for land illegally taken on lease for thirty years from the local Kazaks in 1900, leased it out to other Kazaks and peasants at even higher prices. Finally, by these kinds of operations the settlers drove even some local bureaucrats to despair. They suggested that settlers should be previously selected before coming to Turkestan. Angered by the settlers, who considered themselves a privileged group with their "endless demands, complaints and reproaches," the authorities were ready to do anything in order to drive them out.[72]

State subsidies were believed to further corrupt the settlers who "showed a great carelessness in the struggle for existence and a rare persistence in asking for state subsidies and free-of-charge land." There were also numerous persons among them who did not pay back the received subsidies, leaving their allotted plots before the expiration of the ten-year payment term, as well as those who had been granted subsidies, but in fact did not exist. The wide spread of venereal diseases, especially syphilis, and alcoholism along with a very poor education level became more and more typical. Only a few settlers intended to stay permanently and keep land as their private property.[73] In striking contrast, as ordered by the members of resettlement offices, the local Kazaks often helped the settlers - "the

---

[72] Pahlen, K. K.: *Pereselencheskoe delo v Turkestane*, pp. 14-15.
[73] Ibid., pp. 162-163.

beloved children of the tsar" - with the building of their houses, sowing, as well as providing them with cattle, as they themselves confessed to Pahlen.[74]

To complete the picture, Pahlen cited a curious story told to him by a local police officer during his visit to the Chu River area in Semirech'e. The story was about seven settler families who after receiving state subsidies previously intended for eighty families along with an additional amount of money, which enabled them to return to Russia and bring their relatives with them, established a settlement on their own. In addition, the local Kazaks were persuaded by the policeman to also provide "the guests of the tsar" with portions of their agricultural lands, as well as large numbers of their livestock, including sheep, cows, horses and goats. However, after the Kazaks, as usual, moved off to their summer pastures, the settlers built up pretty houses and went to a local market every weekend to drink away their subsidies, spreading there stories about the infertile soil of their region and the attacks of the local Kazaks. These stories were to prevent possible newcomers from coming to the region, so that these seven families as before could keep in their possession the plots previously allotted to eighty families. After spending all their state subsidies and selling all the cattle given them by the Kazaks, they drank away any money received as well. When the Kazaks returned, they found out that all their millet crops had been reaped and threshed by the settlers. The next year, the seven families maintained that all the land already belonged to them, and the Kazaks must leave their winter pastures or pay for their use: "One group of the good-natured sons of the steppe who used to accept Europeans as their masters obeyed the conditions of the colonists. Those who did not accept them left for the northern areas to till new lands." Having leased their lands that in Pahlen's words resembled a big estate to

---

[74] Pahlen, K. K.: *Im Auftrag des Zaren in Turkestan 1908-1909*, pp. 292-293.

the local Kazaks, these seven families in the end became noted for their hard drinking.[75]

In Pahlen's view, the adherence of the majority of settlers to communal land holding was the main reason for the backwardness of their farming, as well as the underdevelopment of their handicrafts.[76] Yet others (Tresviatskii) considered farmstead land tenure as the reason,[77] though, generally, farm agriculture demanded more efforts from the bureaucrats of local resettlement offices and hence was far more expensive, and generally not promoted by the state. In fact, it was allowed only in completed (*zapolnennyi*) settlements and on the agreement of no less than two-thirds of the members of peasant communities.[78]

None the less, a prominent Kazak leader of the beginning of the last century, Älikhan Bökeikhan, after visiting settlements of both types in Tobol'sk province in 1907, pointed to the much more advanced and prosperous households of those who had practiced farmstead land tenure - mostly the Estonians, the Lets, the Belorussians, the Germans, and the Poles - who had immigrated from Russia's south-western regions and distinguished themselves by their agricultural skills and knowledge in contrast to those whose farming was based on communal land holding.[79]

In any event, regardless of the instructions of guides published for emigrating peasants about the "peaceful and gentle character" of the Kazaks with whom the Russians, as a rule, "get along well,"[80] land occupation unavoidably ran up against Kazak resistance. As reported in 1906 by the newspaper *Rech'*, the organ of the Russian Constitutional-Democratic party, in May

---

[75] Ibid., pp. 334-336.
[76] Pahlen, K. K.: *Pereselencheskoe delo v Turkestane*, pp. 148-150.
[77] *Materialy po zemel'nomu ustroistvu v Aziatskoi Rossii, vyp. 1, Stepnoi krai*, p. 101.
[78] *Zhurnal Soveshchaniia o poriadke kolonizatsii Semirechenskoi oblasti*, p. 32.
[79] Bökeikhan, Ä.: *Tandamaly/Izbrannoe*, pp. 237-241.
[80] Los'-Roshkovskii, F.: *Khodokam i pereselentsam, napravliaiushchimsia v Kustanaiskii uezd Turgaiskoi oblasti (Nastavleniia, opisaniia raiona i uchastkov)*, Poltava, 1912, p. 5.

alone there were at least twenty skirmishes between about five hundred Russian settlers and Kazaks of Turgai province. After hopeless attempts to find free lands, they occupied the lands of the Kazaks, regarding it as state property and built their houses on it. An official declaration of the absence of free lands had not prevented these settlers from coming to the region. Some of them confessed that they preferred to struggle for free land rather than return to Russia and death by starvation there.[81]

In contrast to the official propaganda, Bökeikhan alleged, Russian settlers had brought onto the steppe neither civilization nor wealth but only vandalism and arbitrariness: with a greater awareness of the Kazak mentality, they would capture the Kazak cattle under the pretence of their having trampled their crops and demand unbelievably high redemption money. However, if a peasant lost his cattle, Russians would go to the next Kazak *auyl* and beat all its inhabitants unmercifully, so that it could even result in manslaughter; or they would take Kazaks back with them to their own villages to carry out mob lynching there. Because of the Russian language impediment and the prejudice of bureaucrats at local administrations and courts, innumerable Kazak complaints were simply not taken into consideration.[82]

Among other things, as observed by contemporaries, "a free son of the steppe cannot adapt to borders being sacred. He has got used to thinking that grass growing on the steppe belongs to all grazing their cattle there." Allowing the cattle of settlers to graze on their pastures, they, however, did not prevent their own cattle from encroaching on the settlers' arable lands.

But in such cases, the peasants did not miss the opportunity to immediately fine the Kazaks.[83] The last, as observed by Bökeikhan, recognized only natural borders such as hills, meadows, rivers, lakes, etc., whereas

---

[81] *Rech'*, no. 112 (June, 9 (July 12), 1906).

[82] Kirgiz-Kaisak [Ä. Bökeikhan]: Sud'ba kirgizskogo zemledeliia, in: *V mire musul'manstva*, no. 27 (October, 21 (November 3), 1911.

[83] Kaufman, A. A.: *K voprosu o russkoi kolonizatsii Turkestanskogo kraia*, pp. 31-32.

the former considered as such only the plowed furrow. To avoid an open collision, Kazaks often preferred to roam off. However, in some places the skirmishes became unavoidable: such as one fight that broke out in the Petropavlovsk uezd in which several men on both sides were murdered. Peasants abused their powers, capturing Kazak cattle grazing on the borders with their arable lands. In revenge, Kazaks sowed corn around peasant lands and stole their cattle. As before, the Kazaks considered the whole steppe their property and grazed their cattle almost everywhere.[84] A close neighborhood with the Russian settlements, as observed by a member of the Shcherbina expedition, usually resulted in outrages on the part of their inhabitants, who "damage and mow clean Kazak meadows, fell woods, burn their winter houses and threaten them with axe or scythe in the case of the slightest protest from a Kazak."[85]

All things considered, Pahlen in his final report to the tsar suggested that private land ownership in Turkestan be promoted, by sending instead of the "weak elements of the Russian peasantry and the dregs of the Siberian colonization the strong representatives of the Russian people", who owing to their energy and enterprise would secure the domination of Russian trade and industry there. A forced russification of the region, he warned, would only further aggravate tensions between the Russian and native populations.[86]

§ 3 Colonial Land Policy

Up to 1917 the Cossacks preserved their privileged positions in all matters concerning colonial legislation.

---

[84] Bukeikhanov, A.: Kirgizy, in: *Formy natsional'nogo dvizheniia. Avstro-Vengriia. Rossiia. Germaniia*, ed. A. I. Kostelianskii, SPb., 1910, p. 587.

[85] *Materialy po kirgizskomu zemlepol'zovaniiu, vol. XII: Akmolinskaia oblast', Petropavlovskii uezd*, Chernigov, 1908, p. 48.

[86] Diakin, V. S.: *Natsional'nyi vopros vo vnutrennei politike tsarisma (XIX – nachalo XX vv)*, ed. I. V. Lukoianov, SPb., 1998, pp. 883, 953.

At the earlier stage of their colonization Russian tsars had granted them charters that intended to include the Cossacks as their subjects and confirm their ownership of lands they had already occupied. Later on Cossack land legislation was to be placed under the jurisdiction of Cossack administrations. In more disputable cases, involving as a rule local Kazak and peasant communities, the central ministries could intervene, and among them especially the Ministry of War, to which the Cossacks were directly subordinated. In a word, Cossack legal life was regulated by orders, decrees and instructions issued by imperial as well as their own authorities, and local colonial administrations, with the exception of military matters, had little chance of asserting power over the latter's decisions.[87]

Although unlike the Cossacks', peasant land legislation was supposed to have been placed on a general colonial footing, but in the end it also assumed a kind of "natural law" in character.

As spelled out by the "Regulations on Migrations" of April 8, 1843, one of the first rules aimed at the general regulation of peasant migration, land-hungry (less than five desiatinas) state (free) peasants could move to the regions where they could be granted eight to fifteen desiatinas per male capita.[88] Subsequently, with the declaration of Kazak land state property given to the Kazaks for collective use by Article 210 of the 1868 Provisional Statute (*Vremennoe polozhenie ob upravlenii v Ural'skoi, Turgaiskoi, Akmolinskoi i Semipalatinskoi oblastiakh*), the four steppe provinces received a status equal to the rest of the empire. Accordingly, each settler was to get *gratis* a place for building houses as well as being provided by the local authorities with woods and plots taken away from the indigenous population for agriculture, industry, trade and handicrafts. While the Russian settlers were to be granted the privileges of those registered with the citizens of

---

[87] *Materialy po istorii politicheskogo stroia Kazachstana*, pp. 387-388.
[88] Bekmakhanova, N. E.: *Mnogonatsional'noe naselenie Kazakhstana i Kirgizii v epokhu kapitalizma*, p. 92.

Siberian towns, the Bashkirs and the Tatars, as well as other Asian settlers who had already settled on the steppe were to be deprived of these privileges. Those Kazaks willing to engage in agriculture were to be granted special plots, which along with the buildings built up on them, could be inherited by them, unless the buildings were destroyed and the land no longer used for agriculture. Regarded as private property, these buildings could be sold to others. These measures were to be regarded as furthering the sedentarization of nomadic Kazaks.[89]

While depriving Turkestan's nomadic population of land ownership rights in a similar way, the subsequent 1886 Statute (*Polozhenie ob upravlenii Turkestanskim kraem 1886 goda*) promulgated for the newly conquered regions in Turkestan, granted those rights to its sedentary populations, who could inherit their traditional lands only after they had been surveyed and awarded ownership. This was to be confirmed by special certificates that were to be given to them by local authorities. All "free" lands, including wild forests, were to be considered state property. Non-Christian populations, with the exception of the locals, along with foreigners were prohibited from buying land and buildings in Turkestan.[90]

It was, however, Article 120 of the 1891 Steppe Statute that was to lay down a legal basis for the withdrawal of the so-called *izlishki* (surpluses or excessive lands) in the steppe regions. As formulated by this article, lands remaining after the determination of the amounts of lands used by the local populations were to be considered excessive and along with all wild forests could be withdrawn in favor of immigrating peasants. However, until the ascertainment of their land numbers the nomads could lease them exclusively to Russians for a period not extending thirty years.[91]

---

[89] *Materialy po istorii politicheskogo stroia Kazachstana*, pp. 337-338.
[90] Ibid., pp. 372-373.
[91] Ibid., pp. 395-396.

Yet it was precisely the determination of how much of the land in the possession of the nomadic and sedentary populations in the steppe and oasis regions to the south would be considered necessary, on the one hand, and hence "excessive" or "free," on the other, that became the most tricky point in the whole issue. As L. N. Mordvinov rightly argued, as used in accordance with "custom and tradition," the inherited and tilled lands of Turkestan's sedentary population included all irrigated lands located in its oases, whereas those of its nomadic population included the rest of the surrounding non-irrigated lands. Hence, from the point of view of "custom and tradition," there were no free lands in Turkestan,[92] and correspondingly all those that were to be regarded as "excessive" in the steppe regions.

It should therefore come as no surprise that, formulated in this ill-fashioned manner, colonial land legislation turned in practice into administrative arbitrary rule, where any interference with established local land relations not only proved illegal but also hampered the entire immigration campaign. Accordingly, left to their own devices, local resettlement bureaucrats and administrators were to bear all the burdens of immigration regulation, taking as a rule short-term measures that were to satisfy the settlers' immediate needs. Count Pahlen, who as a just, strict and incorruptible administrator was sent to Turkestan by the tsar in 1908, came to this important conclusion after his thorough review of the local colonial administrations.[93] Though, unfortunately for us, we do not possess an account of similar volume, versatility and value relating to the steppe regions, one might guess that what was known as resettlement policy in those regions fell into the same pattern described by Pahlen for Turkestan.

At the initial stage of colonization, not only resettlement patterns but also those issues generally associated with law and order were often reshaped by the ideas of Turkestan's conquerors - the army generals, since the

---

[92] Mordvinov, N. L.: Zemlevladenie i podati v Turkestane, in: *Tsarskaia kolonizatsiia v Kazakhstane*, p. 265.

[93] Pahlen, K. K. *Pereselencheskoe delo v Turkestane*, pp. 176, 190.

state was reluctant to interfere with their ruling system. The most remarkable of them, Konstantin von Kaufman, the first Governor-General of Turkestan, who "like Cortes during the conquest of Mexico" conquered Turkestan with a small number of soldiers, also came to be known as the author of the first resettlement rules promulgated for the Syr Darya province that later on were applied to the whole of Turkestan. In accordance with these rules, the settlers were to be registered with the citizens of the town Aulie-Ata (Jambyl/modern Taraz). Since 1879, they were also to be granted allowances, the amounts of which as well as the rules themselves constantly changed and underwent partial revisions, depending on the settlers' immediate needs and well-being.[94]

As mentioned, the resettlement of Semirech'e was initiated by its first Governor-General, G. A. Kolpakovskii, who represented an energetic type of a colonizer. Known among the local population as "the general with iron hips", because of his habit of riding long hours on horseback, he would, without being noticed, get up early in the morning, take some milk and bread with him and ride hundreds of kilometers accompanied only by a single Cossack to advise his subordinates. Afterwards he would again ride hundreds of kilometers back to hold important meetings. Born to a poor family in the Russian province of Voronezh with an illiterate father and a peasant mother, Kolpakovskii, thanks to his diligent and long service, rose from an ordinary soldier to an officer of the Russian Army. It was the successful conquest of the Central Asian towns of Pishpek and Aulie-Ata under his command during which he was said to have defeated the much larger Kokand army with only a dozen troops that brought him to the post first of the governor-general of the Semirech'e province and later to the post of the Steppe governor-general.

Considering the colonization of Semirech'e one of his most important duties, Kolpakovskii took on this task with its attendant responsibility. Late in the 1860s, he

---

[94] Ibid., pp. 168-169.

personally visited the peasants of his native province of Voronezh in Russia and after returning to Semirech'e, prepared plots for them. He had also previously met representatives of the local population and enlisted their support, preventing the rise of possible future conflicts. Being inspired by Semirech'e's extraordinarily fertile soil, he was said to have initiated the organization of local agricultural schools in which future colonizers were to learn new methods of farming, including bee keeping and silkworm breeding. He was also believed to have initiated the organization of cotton growing schools and experimental gardens with the fruit samples coming from Tyrol, Germany and France. Presumably, the first peasant settlements, a sort of Potemkin's villages on the Semirech'e soil, were meant to realize Kolpakovskii's own dream of model farming unrealizable in his native Voronezh province. These villages continued to make a pleasant impression with their big gardens and tidy houses even in 1908-1909 when Pahlen first visited them.[95]

Another "iron general," A. N. Kuropatkin, whose "iron diligence, self-control and faithfulness" in the post of the governor-general of Turkestan also helped him to rise to the rank of the Minister of War, adhered to arbitrary principles of rule and treated his subordinates not in accordance with the rules of his profession, but according to their personal allegiance to him. Even after becoming Minister of War, Kuropatkin continued to patronize his subordinates in Turkestan, many of whom were said to be bad-tempered and allowed to abuse their powers and considerably enrich themselves. In Pahlen's words, card playing, women and alcoholism were their main occupations.[96]

Be that as it may, the resettlement of Semirech'e had been conducted in accordance with the private order promulgated by K. von Kaufman in 1881 for the settlement Nikol'skoe until this order was replaced in 1883

---

[95] Pahlen. C.: *Im Auftrag des Zaren in Turkestan 1908-1909*, pp. 213-214.

[96] Pahlen. C.: *Im Auftrag des Zaren in Turkestan 1908-1909*, pp. 271-272.

by the new resettlement rules introduced by the new Governor-General of Semirech'e, Cherniaev, in accordance with which only Russians could settle in the province and in ten years turn their allotted plots into private property. After being in force for only one year, these rules were to be replaced by the new ones drafted by a special commission set up in 1883 on Cherniaev's initiative. In the same vein, however, the commission was abolished already in two years, without arriving at any practical results. Following Kolpakovski's lead, the new Governor-General of Turkestan, N. I. Grodekov, sent a letter to the governor-general of Voronezh in 1884, in which he invited peasants from this province to Turkestan and suggested that they be freed of all taxes for fifteen years. The Voronezh governor-general, in turn, forwarded his letter to the Minister of Inner Affairs, who decidedly came out against all private invitations without the preliminary agreement of his ministry. Yet, despite his prohibition, by the close of 1886 fourteen new Russian settlements totaling 8,400 families had been established in Semirech'e.[97]

The subsequent closure of the province to immigration in 1892 marked the beginning of the unequal struggle of the Semirech'e administration with unauthorized settlers. The first stage was soon lost, and in the same year the local administration was compelled to establish two new settlements - Georgievskoe and Ivanovskoe for more than 1,700 new families. The struggle's next stage began with the promulgation of a new immigration ban in 1895, but was lost again, because local bureaucrats proved to be unable either to stop the stream of immigrants or to chase them out: even the destruction of their settlements proved useless. Finally, in 1902 by the order of Turkestan's governor-general two new settlements, namely Perovskoe and Novo-Nikolaevskoe were established in the province.[98]

As for Turkestan's other provinces, by the 1903 Rules (*Pravila o dobrovol'nom pereselenii sel'skikh obyva-*

---

[97] Pahlen, K. K. *Pereselencheskoe delo v Turkestane*, pp. 172-173.
[98] *Zhurnal Soveshaniia o poriadke kolonizatsii Semerechenskoi oblasti*, pp. 4-6.

telei i meshchan na kazennye zemli v oblasti Syr-Daryinskuiu, Ferganskuiu i Samarkandskuiu) the Russian population and retired low-ranking servicemen of the Russian army in Turkestan were allowed to immigrate and get no more than five desiatinas of irrigated lands per male for perpetual and permanent use. For the first five years the settlers were released from monetary and land taxes (*denezhnaia i zemel'naia povinnost'*) as well as from military service for six years.[99] Meanwhile, by the Rules of June 13, 1903, on the suggestion of the Department of Agriculture and State Properties, the Steppe and Turkestani governor-generals were allowed to withdraw land in the provinces of Akmolinsk, Semipalatinsk and Semirech'e for both the allotment of settlers' plots and the creation of "lands of the country", or the so-called *obrochnaia stat'ia*, and as such could be put on lease by the state.[100]

While declared state property, wild forests in the steppe regions were transformed into the so-called *lesnye dachi edinstvennogo vladeniia kazny*, or forest cottages for the preferential use of the state. In accordance with Article 239 of the 1868 Statute, they could be used "for the allotment of plots for Russian settlements and Kirgiz communities for or without payment."[101] Practical realization of this Article had been carried out on the basis of Instructions number five and six of the Ministry of State Property of May 19, 1898, and May 22, 1899, in accordance with which the choice of territories for *lesnye dachi* was to be left to the local administrations' own discretion. "Land plots baking like pancakes", in the words of the peasant deputy of the First Russian Duma, T. Sedel'nikov, these included as *lesnye dachi* not only forests but also territories with

---

[99] Pahlen, K. K.: *Pereselencheskoe delo v Turkestane*, p. 180.

[100] To *obrochnye stat'i* belonged free lands, fisheries, forests, mills, enterprises, all kinds of buildings, etc., that could be sold or leased by private persons or peasant communities and were to be considered state property; see: *Sbornik zakonov i rasporiazhenii po pereselencheskomu delu i pozemel'nomu ustroistvu v guberniiakh i oblastiakh Aziatskoi Rossii (po 1 avgusta 1909 g.)*, SPb., 1909, p. 532.

[101] *Materialy po istorii politicheskogo stroia Kazachstana*, pp. 338-339.

very poor vegetation, sometimes with four to five trees scattered in various places, or even without any vegetation at all. Accordingly, a forester in the Ural'sk province, without either making any previous agreements with the local population or undertaking any land measurements, simply declared places he liked *lesnye dachi*, such as, for example, the quicksand regions of the Temir district used by the local Kazaks as their winter pastures. The territories of *lesnye dachi* could also be leased out for six years to both the Russian peasants and the Kazaks, though in the case of the former, as Sedel'nikov remarked, they often proved completely unfit for farming.[102]

Another form of land withdrawal widely practiced in the steppe provinces was the so-called *udel'nye* and *kabinetnye zemli* (appanage and cabinet lands) that were to be considered the royal family's private property. Without any regulations, rules or statutes, this kind of land withdrawal was carried out by administrative means.

For instance, when 45,000 desiatinas of important pastures for several thousands of local Kazaks in the Altai region were declared as cabinet lands in 1855, the West-Siberian authorities waged a relentless struggle against them. After peasants were officially allowed to resettle on these lands in 1865, the Kazaks who found themselves living on the territory of the tsar and positively not knowing where to go, drove even the local authorities to despair. The latter's arguments in favor of developing farming among the local Kazaks were not even taken into consideration. The cabinet declared that it had been dealing first of all with the colonization of the Altai region by Russians, who "have been irrepressibly striving to the East and had already established five densely populated volosts." As to the *inorodtsy*,[103] the cabinet was about to "leave

---

[102] Sedel'nikov, T. (deputat I Gosdumy): *Bor'ba za zemliu v kirgizskoi stepi (Kirgizskii vopros i kolonizatsionnaia politika pravitel'stva)*, SPb., 1907, pp. 51- 53.

[103] The term *inorodtsy* (aliens) was originally introduced to designate the legal status granted to the empire's nomadic and hunter-gatherer populations. However, by the turn of the nineteenth century the

them the most insignificant territories". By the order of July 14, 1875, of the Governor-General of Western Siberia Kaznakov, 555 Kazak families were to be evicted from the right bank of the River Bukhtarma that was considered the cabinet's territories, and, in general, from anywhere where, in his opinion, they should not stay. At the same time, the local authorities were ordered not to let these Kazaks to flee to neighboring China. The eviction of these Kazak families was carried out by the peasants who in their anger evicted more than one hundred families at once: they burnt Kazak winter houses, beat and captured the owners themselves if they did not pay them off with money and things. Though a subsequent investigation proved that the disputable land was for the most part unsuitable for farming, Kaznakov insisted on the eviction of these Kazak families, who, finally, fled to China. However, after finding themselves in an even more difficult situation there, they were forced to return to their former places. At least 5,000 of them were then forced to huddle together on a narrow strip of the left bank of Bukhtarma that inevitably resulted in their dramatic impoverishment. In the end, these 555 families, despite all regulations, orders, demands and threats on the part of the local administrators, continued to use their traditional pastures on the right bank of the river even in 1892, nine years after their official eviction. In response, the authorities established a Cossack settlement on the Chingistau Steppe in the Altai in 1882, and the Altai Kazaks, finally, were compelled to commit themselves to maintaining an outpost in Chingistau, so as not to lose these pastures forever.[104]

The cabinet lands of the Kulunda Steppe in Akmolinsk province, which totaled 913,000 desiatinas, had also been leased out to the local Kazaks, and their annual rent reached 3,000 rubles. The figure for the Kazak

---

meaning of the term was extended to refer to all non-Russian natives, see: Slocum, John W.: Who, and When, Were the *Inorodtsy*? The Evolution of the Category of "Aliens" in Imperial Russia, in: *The Russian Review* 57 (April 1998), pp. 173-190

[104] Shmurlo, E.: Russkie i kirgizy v doline Verkhnei Bukhtarmy, pp. 41-62.

families who in 1907 took on the lease of 167,000 desiatinas of cabinet lands in the Kulunda Steppe was 1,089. The remaining 746,000 desiatinas was allotted to Russian peasants.[105]

Subsequently, a 1906 decree granted the Russian nobility and gentry the right to "settle on the state lands of all three provinces and *guberniias* of Asiatic Russia opened for settlers" and take land into their private properties.[106] Some of them such as Princes Ignat'ev, Bezborod'ko, and Iusupov were in the possession of especially large land properties. To Iusupov alone the Kazaks of the Bökei Horde[107] paid 970 rubles annually, so that by the end of the nineteenth century the total of their rent reached 125,000 rubles.[108]

Finally, the construction of various bureaucratic bodies such as *okruzhnye prikazy* (district administration offices),[109] the Orenburg Frontier Commission, etc., but also of enterprises, factories and plants also led to the withdrawal of considerable portions of land. By the 1890s, more than 95,000 desiatinas of land in Turgai province were withdrawn for the building of post roads, railways and stud farms.[110]

The construction of the Siberian railway lines in the 1890s and especially the Orenburg-Tashkent Railway in 1903 further intensified the colonization of Siberian and Central Asian regions. It has been estimated that in the period 1891-1914 five million Russians, Ukrainians and Belorussians settled in Siberia. A special body, the

---

[105] *Zapiska Predsedatelia Soveta Ministrov i Glavnoupravliaiushchego Zemledeliem i Zemleustroistvom o poezdke v Sibir' i Povolzh'e v 1910 godu. Prilozhenie k vsepoddaneishemu dokladu*, SPb., 1910, p. 84.

[106] *Sbornik zakonov i rasporiazhenii*, p. 38

[107] The Bökei Horde was established in 1801 between the lower reaches of the Volga and the Ural Rivers at the request of the Kazak Khan Jängir.

[108] Shonauly, E.: *Jer tagdury – el tagdyry*, p. 65.

[109] In the course of the nineteenth century altogether seven *okruzhnye prikazy* were established in the steppe territories: Kokchetavskii, Karkaralinskii, Akmolinskii, Aiaguzskii, Baian-Aul'skii, Uch-Bulakskii and Aman-Karagaiskii, see: *Materialy po istorii politicheskogo stroia Kazachstana*, pp. 91-202.

[110] Shonauly, T.: *Jer tagdury – el tagdyry*, p. 139.

Committee of the Siberian Railway set up by Sergei Witte in 1892, was committed to the resettlement of peasant settlers along the railway. It withdrew more than 18 million acres (or 7,290,000 hectares) from the pasture territories of the Akmolinsk Kazaks in favor of peasants.[111]

Being first and foremost preoccupied with the withdrawal of as large amounts of land as possible, numerous so-called provisional detachments (*vremennye partii*)[112] urgently set up for this purpose not only neglected the interests of local populations but also the quality of the land they withdrew. As pointed out by Sedel'nikov, the provisional detachments organized in 1893 by the Committee of the Siberian Railway had allotted during the period 1892-1895 eighty-five plots of land totaling 427,431 desiatinas to the Slavic settlers not near the railway, as was envisaged by the Committee, but in the remote uezds of the province - Omsk, Petropavlovsk and Kokchetav - which were lands used by many indigenous peoples.[113] All in all, they were to withdraw eleven million *desiatinas* of the best lands of the region in favor of 700,000 to 800,000 settlers.[114] The building of the Siberian railway, as Sedel'nikov rightly contended, was therefore automatically connected to the colonization of the Siberian provinces of Tobol'sk, Tomsk, Eniseisk and Irkutsk: "It was a simple oversight, but it was considered a law, though it contradicted the Regulations of the Gossovet."[115] Earlier, the West Siberian Resettlement Office established in 1885 allotted something like 33,000

---

[111] Marks, Steven, G.: *Road to Power. The Trans-Siberian Railroad and the Colonization of Asian Russia 1850-1917*, pp, 155, 162.

[112] Provisional detachments for the northern and southern parts of the Akmolinsk province were organized in 1893 and 1900, for the Turgai province in 1893 and 1898, for the Semipalatinsk province in 1898 and 1900, for the Uralsk province in 1900 and 1904 and, finally, for the Semirech'e province in 1904 and 1905; see: Sedel'nikov: *Bor'ba za zemliu v kirgizskoi stepi*, p. 4.

[113] Bekmakhanova, N. E.: *Mnogonatsional'noe naselenie Kazakhstana i Kirgizii v epokhu kapitalizma*, p. 112.

[114] Suleimenov, B. S./ Basin, V. Ia.: *Kazakhstan v sostave Rossii v XVIII - nachale XX veka*, Alma-Ata, 1981, p. 140.

[115] Sedel'nikov. T.: *Bor'ba za zemliu v kirgizskoi stepi*, p. 42.

desiatinas for the settlers of Semipalatinsk province, and for those resettled in Akmolinsk province 280,000 desiatinas.

After visiting Akmolinsk province in 1890, Bökeikhan reported that more than 17,000 settlers had flooded all local Cossack and peasant settlements. Those who could not find either land or places to live had been wandering from one settlement to another, looking for accommodation. All the same, having proven unable to either resettle these peasants or to persuade them not to settle anywhere they wished, the local authorities were left to the mercy of a resettlement bureaucrat who was urgently sent to the province with the task of resettling no fewer than 11,000 settlers in two to three months. A year later, at the request of the Steppe governor-general, a West-Siberian resettlement troop urgently sent to the Akmolinsk and Semipalatinsk provinces allotted about 283,000 desiatinas for twenty-nine peasant settlements to be set up in these provinces.[116] Yet in the end all the allotted land proved unfit for farming, causing the settlers to leave for better plots or return to Russia.

Having thus proved a complete failure, the ill-established work of the provisional detachments finally prompted the government to reconsider its resettlement policy by organizing statistical expeditions, the main task of which consisted in the determination of the so-called nomadic norms and the corresponding amounts of excessive lands relating to particular regions and based on those regions' natural-historical and economic-statistical investigations.

This being the case, a statistical expedition under the leadership of F. Shcherbina between 1896 and 1903 had investigated twelve uezds of the provinces of Akmolinsk and Semipalatinsk, and the two northern uezds of the Turgai province. About forty members of the expedition were divided into three sections and six subsections: the sections were headed by the experienced statisticians, E. V. Dobrovol'skii, V. A. Vladi-

---

[116] Bökeikhan, A.: *Tandamaly/Izbrannoe*, p. 57.

mirskii and L. K. Chermak, and the subsections by P. A. Vasil'ev, Älikhan Bökeikhan, and others.
Unique in its methods, the work of the expedition represented in fact the first attempt at a scientific investigation of Kazak social and economic organization. It served as the basis for several similar expeditions conducted later in the area. As represented by *narodniki,* the adherents of the populist current among the Russian intelligentsia which emerged in the second half of the nineteenth century, the leading members of the expedition were determined to prove the existence of a Kazak *obshchina*. The latter, regarded as an undeveloped form of the Russian *obshchina*, was believed to gradually evolve into the Russian *obshchina*. As is generally known, the *narodniki* idealized the *obshchina* social relations, considering them, among other things, a preparatory step in the transition of Russian society to socialism and as such helping it to avoid the dark sides of capitalist development. After having prepared themselves for their "going native" investigation through a thorough study of the literature about the Steppe *krai*, the Kazaks and the nomadic peoples in general, the members drew up their future investigation program.
Shcherbina proceeded from the assumption that large tribal groups had evolved from a "purely tribal institution" characterized by the unrestricted use of steppe territories to a "purely economic institution" – *obshchina* the members of which grazed their livestock within certain fixed territories (or the so-called *estestvenno-istoricheskie raiony,* natural-historical areas). Accordingly, the members of the expedition termed the Kazak *obshchina*'s basic land tenure unit *khoziaistvennyi aul* (the economic *auyl-* Kaz., a Kazak basic social and economic unit), while the latter's larger confederations *aul'no-obshchinnye gruppy* (*auyl-*community groups) or *zemel'no-obshchinnye gruppy* (land-community groups). Shcherbina considered the discovery of the Kazak *obshchina* the most valuable contribution of his expedition, for the introduced administrative-territorial division did not reflect the real land

tenure pattern practiced by nomads.[117] As this suggested, they began their investigation with the division of the territory of each Kazak uezd into natural-historical areas, or the territories with similar climatic and geographical characteristics, each of which included the traditional Kazak winter (*qystau*), spring (*kökteu*), summer (*jailau*) and autumn (*küzeu*) pastures. The next step in the investigation was the determination of the general figures for both the local populations and their cattle and pastures, including the character of local economies. The last implied the ascertainment of the economic needs of an average local household, including the ways in which they were met. This, in turn, allowed them to calculate the amount of territories necessary for the feeding of a head of livestock (mainly a horse) during each season and the whole year relating to a household, a local community (an economic *auyl*), and a group of communities (*aul'no-obshchinnaia gruppa*) within each natural-historical area. The ascertained figures were to represent the nomadic norms, whose deduction allowed the expedition's members to calculate the amounts of superfluous lands in each natural-historical area and the uezd in general.[118]

Considering the vastness of the steppe territories and harsh climatic conditions accompanied by the absence of either roads or transport, and the organizational and professional difficulties connected, for example, with the general difficulties of a statistical investigation of a nomadic population, it was, indeed, a tough and in many respects unprecedented task that the members of the expedition carried out. The results of their painstaking work were published between 1898 and 1909 in thirteen thick volumes with numerous tables, drawings and maps. According to the expedition's final calculations, from a total of 45,889,000 desiatinas in the territories of the investigated districts, fifty-one percent were to be left to the Kazaks, and the remaining forty-

---

[117] *Materialy po kirgizskomu zemlepol'zovaniiu*, vol. VII: Turgaiskaia oblast'. Aktiubinskii uezd, Voronezh, 1904, pp. 14-16; vol. I: *Akmolinskaia oblast', Kokchetavskii uezd*, Voronezh, 1898, p. III.
[118] *Materialy po kirgizskomu zemlepol'zovaniiu*, vol. VIII: *Semipalatinskaia oblast', Zaisanskii uezd*, SPb., 1909, pp. V-VI.

nine percent were to be used for the resettlement of the Russian immigrants. In 1901, the Ministry of Agriculture and State Property increased the Kazak norms to 29,121,000 desiatinas or sixty-three percent of all their former territories, which represented a twenty-five percent increase from their Shcherbina norm.[119]

However, as critics of the methods applied by the expedition argued, the real character of the Kazak nomadic land tenure system had a much more complicated and intricate character, and the use of land by *auyl*-community groups could not be bound to particular natural-historical areas. As A. A. Kaufman, himself a member of a statistical expedition, showed, some tribal groups completely unrelated to each other could use the same natural-historical area, and, vice versa, one tribal group could graze their cattle on the pastures of two or even three such areas.[120] He went on to say that in reality not the *auyl*-community groups but their smaller components - the *auyls* constituted a basic economic unit determining the character of Kazak land relations. As it turned out, even if the *auyls* grazed their cattle within the same tribal group and natural-historical area, the sizes of their pastures could considerably differ from each other, depending on season, climatic and geographical conditions, as well as the composition of the *ayuls* themselves. Hence, all figures calculated by the expedition for each natural-historical area and relating to both the *auyl*-community groups and excessive lands had no practical sense, for the withdrawal of the latter could theoretically lead to the loss of all pastures of one group of *auyls,* and at the same time to the preservation of all their pastures by others. Moreover, the withdrawn land could also be of different value, and not always fit for farming purposes, for land fertility from the point of view of farming or nomadic cattle breeding was not the same thing. The high productivity of certain pastures turned out to have not only little value in those cases when they were not used by the nomads, as had also

---

[119] RGIA, f. 1276, op. 4, d. 468, l. 200.
[120] Kaufman, A. A.: *K voprosu o russkoi kolonizatsii Turkestanskogo kraia*, p. 6.

been observed by the expedition's members themselves, but also given the changeability of climatic conditions, to be an inconstant quantity in itself.[121] In addition, since the Kazaks often commonly used their pastures, the withdrawal of all lands indicated as superfluous could damage first of all the poor, because the rich, owing to their influence, would be able to pasture their cattle on the diminished territories, whereas the poor would be forced to either drastically reduce their cattle or to roam off.

Another strong point in A. A. Kaufman's criticism related to the neglect by the expedition of Kazak hay mowing lands, playing an ever-growing role in their economy. Similarly, regarding the winter pastures as being divided between different *auyls* and the summer pastures as common property of *auyl*-community groups, they believed that the withdrawal of superfluous lands calculated for the latter's summer pastures would not cause damage to any single *auyl*. In fact, however, winter pastures, depending on the region, could be in the common use of many *auyls*, whereas the summer pastures could have separate owners.[122]

Significantly, refusing in principle the very idea of the determination of nomadic norms for an economy that depended on a number of "such complicated fine demands" from nature that could not be envisaged by any instructions, A. A. Kaufman finally arrived at the conclusion that only after the consultation in each case with the members of particular *auyls* themselves, could the amounts of superfluous lands be determined - the method that had been successfully used by the first Russian settlers.[123]

---

[121] *Materialy po kirgizskomu zemlepol'zovaniiu*, vol. XII, p. 9.

[122] Kaufman, A. A.: *Materialy po voprosu ob organizatsii rabot po obrazovaniiu pesereselencheskikh uchastkov v stepnykh oblastiakh (Iz otcheta starshego proivoditelia rabot Kaufmana po komandirovke v Akmolinskuiu oblast' letom 1897 g.)*, SPb., 1897, pp. 48, 90-91.

[123] Kaufman, A. A.: *K voprosu o russkoi kolonizatsii Turkestanskogo kraia*, pp. 172-173.

As for other critics of the Shcherbina method, both Pahlen and P. Rumiantsev[124] also pointed out that the character of the land tenure of Turkestan's nomadic population had very little to do with the "natural-historical areas" into which Turkestan had been previously divided; and the activities of the local resettlement offices aimed at the distribution of land "*nach einer wissenschaftlichen Formel*" had in the end resulted in excitement and indignation on the part of the natives, so that only the threat of military punishment had kept them from open resistance.[125]

To top it all off, the basic information used by the members of the expedition in their investigation was provided through translators by Kazaks who, as they themselves admitted to A. A. Kaufman, deliberately diminished the figures for their cattle and gave false information concerning the location of their winter and other pastures, so as to avoid any trouble.[126] In addition, inevitable translation difficulties relating to the understanding of basic notions furthermore complicated cooperation. For example, the Kazak traditional linear measure *shaqyrym* that literally designated the distance at which one person could hear the voice of another and consequently varied from one place to another could not be easily transformed into the official unit of measure (verst) adopted by the expedition. Likewise, when the expedition's members tried to ascertain the degree to which certain pastures were used by Kazak cattle, they got different answers from their owners each time: If some Kazaks logically regarded those pasture lands where the grass had been entirely fed upon by their cattle as completely used, the others considered such pastures as places where their horses "stopped to play and gambol" and did not put on

---

[124] Rumiantsev, P.: *Materialy po obsledovaniiu tuzemnogo i russkogo starozhil'cheskogo khoziaistva i zemledeliia v Semirechenskoi oblasti*, vol. 3, SPb., 1912, p. 160.

[125] Pahlen, C.: *Im Auftrag des Zaren in Turkestan 1908-1909*, pp. 288-290.

[126] Kaufman, A. A.: *Materialy po voprosu ob organizatsii rabot po obrazovaniiu pereseselencheskikh uchastkov v stepnykh oblastiakh*, pp. 64-65.

weight, though there could be plenty of grass on them.[127] Similar problems arose with the ascertainment of the borders of those pastures, which, in accordance with the owners' answers, had always been determined by their horses.

Consequently, marked on the expedition's maps as comprising 25,000 desiatinas, free lands of one of the steppe regions, proved after an additional investigation to constitute only 6,000 desiatinas.[128] Analogously, a repeated calculation of the amounts of the excessive lands in Turgai province gave the figure of 1,300,324 desiatinas instead of the previously calculated 1,971,094. Conversely, the figures relating the Kazak families living in the Kustanai uezd of the province increased from 7,756 to 8,649; in other words, 893 families were left without land by the expedition.[129]

As all critics pointed out, apart from the expedition's leading figures, the majority of its members represented invited persons who had a very poor idea about their work as well as the nature of Kazak nomadic economy. More than this, the expedition was under strong time pressure: After having begun its work in the summer of 1896, it had investigated by November only two uezds, although the figures for nomadic norms calculated for all uezds were to be delivered by March of the following year, i. e. by the beginning of the sowing campaign. Yet, the expedition was able to submit the first results of the investigations only by the middle of the following June.

All the more amazing, however, is the fact that though indicating all the difficulties of their investigation pointed out by their critics, the members of the expedition, being under even more urgent pressure to deliver the calculated nomadic norms imposed on them by the Ministry of Agriculture and State Properties, were in many cases forced to generalize their results and overlook some important details. Yet even when their figures considered all conditions relating to both local

---

[127] *Materialy po kirgizskomu zemlepol'zovaniiu*, vol. I, p. 171.

[128] Sedel'nikov, T.: *Bor'ba za zemliu v kirgizskoi stepi*, p. 42.

[129] Shonauly, T.: *Jer tagdury – el tagdyry*, pp. 123-125.

areas and economies and were provided with correct data, as they themselves assured, there was no guarantee that in several years these figures would not need considerable revision. More importantly, the fixing of nomads to certain natural-historical areas proved as ineffective as was their official fixing to administrative-territorial units.

With a new law of June 6, 1904, permitting the Russian peasants from European Russia to emigrate and encouraging them with privileges,[130] as well as a subsequent decree of March 4, 1906, about the organization of provincial and district commissions on land organization (*gubernskie i uezdnye zemleustroitel'nye komissii*), the latter's members, however, did not stick to the norms calculated by the Shcherbina expedition for the steppe provinces, cutting off, as a rule, much bigger territories, or using them for areas with different climatic conditions and local economies. As it is, the nomadic norms calculated for the Kokchetav uezd of Akmolinsk province had been used for other uezds of the province, as well as the provinces of Semipalatinsk and Turgai, though Kokchetav uezd, as far as the fertility of land was concerned, represented the exception to all other steppe regions rather than the norm, according to Bökeikhan, a member of the Shcherbina expedition. He believed that due to the unfavorable climatic conditions - limited precipitation and water sources for both people and cattle, including the peculiar character of the steppe soil which secured high harvests only in the first years and was relatively quickly used up - the steppe regions for the most part were unsuitable for farming: "It is not enough to calculate the numbers of excessive lands in the Kirgiz [Kazak] Steppe. One should think of providing them with water, without which tens and hundreds of thousands of steppe territories will not have any value."[131] All the same, being highly skeptical about the system of the nomadic norms in general, Pahlen remarked that if the nomads of Semirech'e were granted lands two-fold larger than those

---

[130] RGIA, f. 1276, op. 4, d. 468, l. 211.
[131] Bökeikhan, A.: *Tandumaly/Izbrannoe*, pp. 242, 254.

proposed by the nomadic norms, the resettlement bureaucrats would in the end be compelled to take away their tilled lands as well as those worked by the sedentary population, because of the lack of irrigated lands in the province.[132]

Even the method introduced by some resettlement bureaucrats, whereby several infertile desiatinas were counted as one fertile desiatina, could not improve the situation, because "wheat not growing on one infertile desiatina will, obviously, not grow on five such desiatinas."[133] Hence, however large the allotted plots might be, often only one eighth of them proved fertile. Consequently, the resettlement bureaucrats of two uezds of Akmolinsk province were forced to admit that all 1,400 plots allotted during 1906 to 1907 were completely unsuitable for peasant resettlement. Another trick used by resettlement bureaucrats consisted of the following: In order not to pay for the destruction of Kazak winter dwellings left on the territories that were to be withdrawn, they measured and cut off the whole surrounding territories, leaving out the dwellings themselves: they knew that trying to avoid such close neighboring with the peasants, the Kazaks would leave their dwellings on their own.[134]

Various instructions issued for resettlement bureaucrats demanded that wells, roads and graves as well as other important places like such as watering places were to be left to Kazaks. As a rule the Kazaks wished their land to be taken away in compact portions in order to avoid strip farming with the Russian peasants, and therefore regarded as excessive the scraps of infertile land scattered in different corners of their pastures. The same, however, was true for the Russian peasants and the Cossacks who for their part demanded that lands they might need not be included in the allotted plots. Given all these conditions that were principally unrealizable in practice, the activities of the provisional detachments assumed in the end an arbitrary character,

---

[132] Pahlen, K. K: *Pereselencheskoe delo v Turkestane*, p. 61.
[133] Bökeikhan, A.: *Tandamaly/Izbrannoe*, p. 236
[134] Ibid., p. 248.

proving thus the complete failure of a "scientific" approach to the resettlement campaign undertaken by the government.

Pointing out the three and a half million desiatinas that had been withdrawn in the steppe regions during 1892-1906 by the provisional detachments on a completely illegal basis, Sedel'nikov admitted that:

> In fact, a planned and purposeful work on the allotment of settlers' plots has never existed on the steppe. Up until recently all work has been done in a haphazard and slapdash manner, by fits and starts, in one place in one way, in another, in quite a different way, and elsewhere, quite out of place, thus bringing about an even bigger mess into the already complicated land situation on the steppe... In practice all work had been reduced to a desire not to merely satisfy but to amaze the authorities with the quickness and successfulness of the work, without adding any special significance to its quality, expediency and sensibility.[135]

A sad picture appeared before the eyes of Bökeikhan during his visit to the peasant settlements of Tobol'sk province and led him to conclude that their inhabitants had become the victims of resettlement activities: many of them, after finding themselves in a disastrous situation, were forced to sell their houses at very low prices to newcomers and to leave for other places. The settlers refused to pay taxes and perform military duties, maintaining that: "The tsar knows nothing about it, if he knew where we live, he would demand neither taxes nor soldiers." They believed that the local resettlement bureaucrats alone were to blame for the disastrous situation in which they had found themselves.[136]

Analogously, numerous commissions, committees, and other official bodies set up in Turkestan and patterned on the Shcherbina expedition, also failed to achieve their goals. A special commission organized in 1905 at

---

[135] Sedel'nikov, T.: *Bor'ba za zemliu v kirgizskoi stepi*, pp. 39, 49-52.
[136] Bökeikhan, A.: *Tandamaly/Izbrannoe*, p. 241.

the persistent request of Turkestan's governor-general that included the representatives from the Ministries of Agriculture and State Properties investigated only one volost and was already disbanded in 1906. Nevertheless, a detachment set up on a local initiative allotted thirty-six new land plots totaling 288,990 desiatinas for more than 50,000 settlers in various places of Semirech'e between 1905 to 1907.[137] Although the Law of February 14, 1905 addressed the creation of settlers' plots and demanded the satisfaction of the interests of the Semirech'e's nomadic people first and foremost, the corresponding investigation of the province had never been carried out.[138] After a commission under the leadership of General Korol'kov set up in 1908 for the same purposes also failed to arrive at any practical results, namely the determination of the amount of excessive lands, the governor-general of Turkestan initiated the organization of a new commission that was to provide the settlers with urgently needed medical assistance, food, and transport, as well as plots of land.[139]

In much the same vein, however, even if the statistical expeditions carried out their work, their results proved useless in the end. One statistical expedition in 1907 investigated only nineteen Turkestani uezds; however its results were used for the allotment of plots in 130 other uezds. In many cases the allotment of plots was carried out simultaneously along with statistical investigations or even without them at all. Consequently, the norms determined in such a hurry contained many mistakes and inaccuracies. Not only a lack of professionalism but also of tact in respect to the local populations were characteristic of the work of topographers. They often indicated infertile lands as fertile and vice versa, as well as marked non-existing roads and other objects on their maps, pursuing their only goal, namely to take as much land as possible. In so doing, one resettlement detachment practicing the withdrawal of

---

[137] *Zhurnal Soveshchaniia o poriadke kolonizatsii Semirechenskoi oblasti*, p. 4.
[138] Diakin, V. S.: *Natsional'nyi vopros vo vnutrennei politike tsarizma*, p. 891.
[139] Pahlen, K. K.: *Pereselencheskoe delo v Turkestane*, p. 180

not only "excessive" but also the arable lands of the local population along with the destruction of their buildings, destroyed in 1908 in Semirech'e more than 5,100 local settlements with a population of about 30,000 in order to get something like 25,000 desiatinas of fertile lands for 6,500 Russian settlers' families. However, only 2,300 or 38 percent of those families could be resettled on those lands, because the majority was infertile. Another statistical expedition gave the figure for fertile lands in the Semirech'e province as comprising even a million desiatinas. This served as the basis for the local resettlement office, initiating a large resettlement campaign and requiring considerable state subsidies. However, after another investigation, only 200,000 desiatinas of the previous figure proved to be suitable for resettlement.[140]

As summed up by Pahlen, despite numerous rules and instructions, the resettlement of peasants had been carried out every time based on the basis of private agreements often achieved by means of pressure by the local administrations on the native populations. The latter, in accordance with these agreements, were to leave their lands together with their buildings for certain financial compensation.[141] Similarly, a sometime observer of the land withdrawal procedure in the steppe regions, Sedel'nikov described it in the following way:

> A superintendent of work (*proizvoditel' rabot*, a member of a resettlement detachment), as a rule, a poorly educated though literate man, in addition provided with various kinds of instructions and provisional rules - on the one side, and the ignorant and naive Kirgizs, who felt a particular and mystical respect for all sorts of 'papers' and 'authorities,' on the other. As a result, a 'voluntary agreement,' sometimes accompanied by blackmail and the threat to take away much more

---

[140] Pahlen, K. K.: *Pereselencheskoe delo v Turkestane*, pp. 68, 57, 77-79.
[141] Ibid., p. 183.

of their lands than had been indicated in his papers.[142]

The category *pereselenets* (settler), as Pahlen attested, had no clear meaning and included all those arriving in Turkestan. Independently of whether they had any papers with them or not, all of them were automatically registered as settlers and granted land. Accordingly, along with the citizens of Russian cities one could also find among them also peasants who had been living in Turkestan for more than one hundred years. In 1907, the local authorities of the Pishpek region, under the aggressive pressure of the newly arrived settlers, resetled all of them, without asking for any documents. Finally, even the citizens of both Pishpek town and those emigrating from Russian cities were also resettled and granted land.[143]

Among other things, Pahlen's mission to Turkestan envisaged the settling of a conflict between the local governor-general, P. I. Mishchenko, and the head of the Semirech'e resettlement office, Veletskii. Shkapskii, who objected to the policy of land withdrawal, replaced the latter in 1906.[144] Generally, the conflict was caused by the lack of any coordination, let alone collaboration between local and resettlement authorities, which to no small degree was predetermined by the latter's self-dependent position and their direct subordination to an independent body – the Chief Administration on Land Organization and Agriculture (*Glavnoe Upravlenie Zemleustroistva i Zemledeliia* – the GUZiZ). The first government body charged with resettlement matters was the Resettlement Office. It was established in 1896 and attached to the Ministry of Inner Affairs. Yet, with the rapid growth of emigration followed by the subsequent division of the steppe regions into four resettlement regions (*pereselencheskie raiony*) in 1904: Uralo-Turgaiskii, Akmolinskii, Semipalatinskii and Syr Daryinskii, and the establishment of local resettlement offi-

---

[142] Sedel'nikov, T.: *Bor'ba za zemliu v kirgizskoi stepi*, p. 49.

[143] Pahlen, K. K.: *Pereselencheskoe delo v Turkestane*, pp. 85-88.

[144] Qoigeldiev, M.: *Alash qozgqlysy*, p. 36.

ces in each, the need for an independent organ for resettlement matters became obvious.

The Resettlement Office was placed under the Ministry of Agriculture and State Properties that was renamed in 1905 the Chief Administration on Land Organization and Agriculture.

In substance, the conflict between the local authorities only further aggravated the land situation in the region, the dramatic consequences of which were to affect both the Russian and indigenous populations. In addition, the lack of any competence in matters concerning Kazak land organization made the local steppe administrations completely useless in general for the purposes of the resettlement campaign, according to Sedel'-nikov.[145]

Yet Mishchenko insisted on the subordination of resettlement issues to the local authorities, as they would know the local conditions better. Following the lead of his predecessor, Grodekov, he decidedly came out against the activities of local resettlement detachments, neglecting the vital interests of the local populations by withdrawing their irrigated lands. He argued that only after the natives were provided the lands necessary for leading their traditional economies, could the remaining lands be used for the allotment of plots to peasants. Both governor-generals, Grodekov and Mishchenko, refused to approve the norms of the GUZiZ, characterizing them as "spun out of thin air," and pointing (Mishchenko) to the discontented Kazaks leaving for neighboring China. Grodekov, for his part, also warned about the possible dramatic consequences of the resettlement policy that could lead to the indigenes' resistance. Citing his arguments, the Ministry of War applied to the government in 1908 for the closure of Turkestan to immigration and the creation of a legal basis for land withdrawal. The Ministry also pointed out that without an irrigation system the Turkestani lands, with the exception of Semirech'e, would not have any value. Nevertheless, the chairman of the GUZiZ, B. A. Vasil'chikov, regarded this policy as

---

[145] Sedel'nikov, T.: *Bor'ba za zemliu v kirgizskoi stepi*, pp. 49-50.

promoting the transition of the nomads to a sedentary way of life and not as violating their vital interests. Though admitting that the irrigated lands of the local populations had been withdrawn and their houses taken away, he argued that the determination of nomadic norms would demand an "uncertain lengthy time." In the case of "unlawful solicitations" on the part of Kazaks, however, they were to meet with the "rebuff of the local administrations." What is more, as he put it, "not to actively promote the sedentarization of the Kirgizs [Kazaks] means to neglect the needs of the Russian people."[146] The resettlement bureaucrats with Glinka at their heads blamed the local authorities for their interference in the resettlement matters and their generally "unfriendly attitude to the Russian cause" in the remote outskirts of the empire.[147] As Pahlen remarked, they presented their activities as a "harsh and tenacious struggle for the Russian cause that has been prevented by the local administration's selfish aims."[148] Finally, a new governor-general, A. V. Samsonov, who promoted the government resettlement policy, replaced Mishchenko.

Qualified as doctors, veterinary surgeons, architects, engineers, foresters, bacteriologists, chemists, including also all kinds of naturalists, botanists, statisticians, teachers, land surveyors and agronomists, and privileged by salaries five times larger than those of local administrators, the resettlement bureaucrats quickly became the subject of the latter's hatred. They were irritated by the boasts of their colleagues about their connections with the influential bureaucrats from St. Petersburg, but foremost by their "amateurish and often senseless waste of state subsidies." They sent their "forged reports to St. Petersburg with long columns of figures" which were to demonstrate the large amounts of fertile lands and resettled territories. With the goal of "throwing dust in the eyes of remote St. Petersburg,"

---

[146] Diakin, V. S.: *Natsional'nyi vopros vo vnutrennei politike tsarizma*, pp. 891-895.

[147] Pahlen, K. K.: *Pereselencheskoe delo v Turkestane*, p. 127.

[148] Diakin, V. S.: *Natsional'nyi vopros vo vnutrennei politike tsarizma*, p. 896.

these reports were at the same time, as Pahlen put it, "to cover the true activities of their authors, namely their passion for intrigue and self-seeking."[149]

While visiting a botanical laboratory organized by Veletskii in Vernyi, Pahlen himself became a victim of this policy of "throwing dust." His description of his visit runs as follows:

> At a fixed time Mr. Veletskii with a number of agronomists, chemists and employers expected me at the entrance of the laboratory. Housed in five big and beautifully decorated rooms of a house with large columns, the laboratory was furnished with tables with glasses and mortars on them. High flames were burning under various liquids, and the chemists sitting at the tables seemed zealously at work. Props with the sacks of soil samples hung on the walls. Everything had a semblance of the carrying out of scientific experiments. After looking through the books and reports, I remained for about half an hour till everything was shown to me. In the bureau of the laboratory furnished with American furniture I asked Mr. Veletskii: "Tell me, please, when did you set up the laboratory, two weeks or eight days ago?" The faces of those present fell, and that of Mr. Veletskii turned pale. The chief of the laboratory first tried to evade the question, but then confessed that they had built the laboratory shortly before my arrival and had had no time to work in it. The figures in the reports were calculated blindfold and therefore contained numerous mistakes. The maps on the walls were brought from St. Petersburg. I began to slowly understand that even the samples of soil were not original. I have had many opportunities in my life to advise good laboratories at agricultural institutions, and have also had some agricultural experience as a former farmer, but I have never seen such a brand new laboratory yet. One could notice that the

---

[149] Pahlen, C.: *Im Auftrag des Zaren in Turkestan 1908-1909*, pp. 274-276.

lamps as well as the glasses and pans have not ever been used. There were no spots either on the tables or the walls. Only the first pages of the books and maps contained the results of chemical and physical investigations registered with the same handwriting and with ink that with due course has not yet even become dark. The trouble was that these chemists had already been paid for many years, during which time they regularly sent their fabricated reports on soil investigations to St. Petersburg.[150]

If, before 1901, the resettlement offices were to withdraw only twenty-five percent of the number of lands indicated as excessive by the Shcherbina expedition, then by the time of the implementation of Stolypin's agrarian reform (1906) all "excessive" lands had been taken away.[151] A GUZiZ meeting held in 1909 did not approve Pahlen's critical remarks nor his suggestions that among other things pointed to the illegal character of the land withdrawal practiced by the local resettlement offices in Turkestan and suggested a suspension of emigration to Semirech'e. Nevertheless, they unanimously found themselves in favor of the continuation of current policy.[152]

All the same, despite constituting the centerpiece of Stolypin's agrarian reform, the implementation of the government emigration campaign did not contribute to its legal regulation and was, as before, conducted by local authorities without any basis in law. Considering the colonization of "the vast, free and useless regions of Asiatic Russia" by a huge bulk of the impoverished Russian peasants one of the most important conditions for the establishment of strong farm agriculture in Russia, Stolypin's government had increased the financial support of the campaign five-fold in the period 1906-1915. The support included credits for building houses (35.6 percent) and roads (14.1 percent), medical

---

[150] Pahlen, C.: *Im Auftrag des Zaren in Turkestan 1908-1909*, pp. 304-306.

[151] Sedel'nikov, T.: *Bor'ba za zemliu v kirgizskoi stepi*, p. 39.

[152] Diakin, V. S.: *Natsional'nyi vopros vo vnutrennei politike tsarizma*, pp. 951-954.

assistance (15 percent), measurement (12.4 percent) and irrigation (6.5 percent) works, the maintenance of resettlement offices (3 percent), and others.[153] In addition, the campaign was accompanied by the publication of numerous pamphlets and guides for emigrating peasants that provided them with necessary information about climatic conditions, prices, names, necessary addresses, and descriptions of landscapes and roads.

The Shcherbina nomadic norms were found too generous, and a new expedition under the leadership of K. V. Kuznetsov was set up in 1907 for the reinvestigation of the regions previously investigated by the Shchebina expedition. Working from 1907 to 1911 on the program of its predecessors, the Kuznetsov expedition cut the norms established by Shcherbina by half. All in all, by 1917 the state had withdrawn forty-five million hectares or sixteen percent of all Kazak lands for various purposes, but basically for the resettlement of settlers.[154]

Apart from being considered an important means to solve Russia's domestic problems, the colonization of remote Asiatic and Siberian regions was to indicate a new approach to matters concerning her international policy, in accordance with the decisions of a government meeting of the Ministry of Internal Affairs held in 1907 under the chairmanship of A. I. Lykoshin. Since the war against Japan was lost because of the scarcity of population and the lack of a proper means of communication, the participants at the meeting urged the fastest possible resettlement of those regions be considered the government's top priority policy. Thereby the resettlement of the nomads was to follow the pattern adopted for the Russian settlers, for according to "the people's sense of justice," giving preference to the interests of the Kazaks "hurts the Russian peasant who has shed his blood in loyalty to the Russian state."[155]

---

[153] *Materialy po zemel'nomu voprosu v Aziatskoi Rossii*, vyp. 6: *Itogi pereselencheskogo dela za Uralom s 1906 po 1915 gg.*, ed. V. A. Tresviatskii, Petrograd, 1918, p. 3.

[154] Qoigeldiev, M.: *Alash qozgalysy*, p. 50.

[155] Diakin, V. S.: *Natsional'nyi vopros vo vnutrennei politike tsarizma*, p. 843.

Yet, almost a third of altogether 3,800,345 *khodoki* who were reported to have crossed the Urals between 1906 and 1915 did not stay, and either returned back to Russia or went to other resettlement places.[156] Among the main reasons that had prevented their staying, as reported by deputy A. L. Tregubov in 1909 to the Third Duma, were bad harvests, harsh climatic conditions, and the growth of voluntary immigration, but also the high registration prices with peasant settlements established earlier, including the prohibition of leasing Kazak lands as well. A lack of religious practice along with the loss of a family or its head could also prompt the peasants to move.[157]

As for the Kazaks, Stolypin and Glinka in the memorandum written after their trip to Siberia and the Povolzh'e region in 1910 unequivocally pointed out:

> The methods directed at the resettlement of the Kirgizs [Kazaks] are mistaken. We are deeply convinced that not the Kirgizs but the steppe itself needs to be settled, and we must think not of individual Kirgizs, but of the future of the whole steppe. From this perspective, the resettlement of the Kirgizs cannot be understood in terms of providing them with all their present territories. Only those of them who are quite prepared to settle can and must be resettled.... As for those vast areas, where the Kirgiz economy has not yet been improved and developed enough, we should continue the policy of the withdrawal of nomadic lands, leaving to the nomads as little land as possible.

The land of the nomadic Kazaks was to be regarded as given to them only for provisional use with a subse-

---

[156] *Materialy po zemel'nomu voprosu v Aziatskoi Rossii*, vyp. 6: *Itogi pereselencheskogo dela za Uralom s 1906 po 1915 gg.*, pp. 1, 6-7.

[157] Tregubov, A. L.: *Pereselencheskoe delo v Semipalatinskoi i Semirechenskoi oblastiakh. Vpechatleniia i zametki chlena Gos. Dumy A. L. Tregubova po poezdke letom 1909 g.*, SPb., 1910, pp. 6-8.

quent expropriation in the interests of the state and the creation on this basis of "the lands of the country."[158]

As for those Kazaks who were "unprepared to settle," the Instructions of June 9, 1909 instructed the land surveyors to provide them with the "amounts of land necessary for leading their traditional way of economy" that were to be defined for each separate group of them. The latter could even be granted additional lands in case of the lack of necessary territories in their areas. Moreover, their winter pastures with all buildings and fences, irrigation buildings and dams, including orchards, kitchen gardens and plantations were to be reserved for them along with their irrigated and arable lands, cemeteries, graves, burials, and mosques as well. Otherwise, the Kazaks were to be paid for the destruction of their winter camp buildings and other material losses.[159] Yet it was precisely the impossibility of determining the "amounts of lands necessary for leading a nomadic economy" that made these instructions useless, like all other instructions and orders promulgated on the subject earlier useless, for none of them made things clearer. Nevertheless, the 1914 law (*Pravila o pereselenii na kazennye zemli*) allowed the peasants to immigrate or send scouts to Asiatic Russia and get plots there for permanent use, despite the fact that "the rules regulating the final resettlement of the indigenous populations have not yet been laid down."[160]

On the other hand, those nomads willing to settle were obliged to give up their nomadic status and allow themselves to be registered with the status of the Russian peasants; i. e. they were to be subordinated to the latter's social and administrative institutions and correspondingly to be granted ten to fifteen desiatinas of land, paying instead of *kibitochnaia*, *obrochnaia* taxes.[161] This latter measure included, among other

---

[158] *Zapiska Predsedatelia Soveta Ministrov i Glavnoupravliaiushchego Zemledeliem i Zemleustroistvom o poezdke v Sibir' i Povolzh'e v 1910 godu*, p. 22.

[159] *Sbornik zakonov i rasporiazhenii*, p. 638.

[160] *Istoriia Kazakhskoi SSR*, vol. 3, Alma-Ata, 1979, p. 410.

[161] *Sbornik zakonov i rasporiazhenii*, p. 638.

things, the forced resettlement of Kazaks next to Russians, a proximity which was to facilitate the Kazaks' conversion to Orthodox Christianity. According to the decision of the Holy Synod of February 5, 1914, the Kazaks were to be resettled together with the Russians and their total number was not to exceed half the number of the Russian population in order to give advantage to the latter in all decisions.[162] Registration as peasants also involved obligatory service in the Russian army, whereas nomadic status released *inorodtsy* from military service.

It is little wonder then that having found themselves facing the unsolvable dilemma of either remaining nomads and in the end being subjected to eviction, or of settling and farming, but under the condition of their transformation into the Russian *muzhiks* (peasants), the Kazaks preferred to keep faith with their accustomed way of life.

The dilemma, however, seemed unlikely to become cause for deep concern to the statesmen, who categorically maintained that the Russian settlers had brought culture, civilization, progress and prosperity into the steppe, promoting the transition of Kazaks to sedentarization. Resolutely rejecting all arguments about the impoverishment of Kazaks caused by the Russian immigration, they affirmed in their memorandum:

> This is a normal trend of development. The Kirgizs cannot remain eternal nomads, unless they are capable of adopting culture. The experience of the last few years has demonstrated that they are capable of farming, and that the Russian immigration connected with an unavoidable limitation of pasture territories has proved a powerful and until recently the only stimulus to their transition to sedentarization. Hence, a jealous protection of the *chernozem* (black earth) of the Kirgiz Steppe and the nomadic economy in general from the Russian farmers would be erroneous in all respects, even with respect to the Kirgizs themselves. It would be especially in-

---

[162] RGIA, f. 391, op., 4, d. 1663.

excusable from the point of view of Russian statehood (*russkoi gosudarstvennosti*) and culture.

In this context, the introduction of the social institutions of the Russian peasants was also expected to smooth over all peculiarities of the nomadic social life, meeting thereby the interests of both the Kazaks and the state. The authors believed that from the point of view of the state's interests, the Kazak Steppe with the help of the Russian peasants would even be able to feed the European markets.

Citing as an example the grassy Kulunda Steppe in Akmolinsk province, where in the space of only two years 200 settlements with 55,000 peasants had been resettled, they enthusiastically asserted: "The pulse of Russian life begins to throb on the steppe that had been dead before; we traveled through the fourteen settlements of the Kulunda Steppe and saw everywhere economic awakening; the steppe is being covered by wells... This awakening of life makes a deep impresssion: we have witnessed how a successful resettlement could increase people's prosperity."

Two developments observed by them on the steppe, namely the transition to sedentarization on the part of some Kazaks, and the moving off to remote districts of other Kazaks were regarded by them as positive phenomena, because in both cases the remaining free lands were very quickly occupied by settlers.

Provided with the forged data of the resettlement offices, the authors wrote about the huge amounts of free fertile lands beyond the Urals, something like twelve million desiatinas in Akmolinsk province alone: "Though losing millions of desiatinas, they [the Kazaks] have been rewarded by the fact that their remaining land, including arable lands, meadows and cattle, have for the first time acquired a market value." They cited the same data to demonstrate both a considerable increase in the number of prosperous Kazaks and a corresponding decrease in the poor of Akmolinsk province between 1898 and 1908. According to these statistics, in some areas of the Omsk district, up to 95-100 percent of Kazak households were engaged in farming. The figures of mowing machines and cattle in Kazak possession, as well as their traditionally high birth rate were also intended to serve as proof of their growing prosperity.[163]

However, when reported from a "native perspective" by the Kazak magazine *Aiqap*, the trip of the above-mentioned statesmen, including also the Minister of Agriculture, A. V. Krivoshein, appeared in a somewhat different light, revealing some important details. Having limited themselves to the visit to only several Russian settlements located not far from the town of Petropavlovsk, according to the newspaper, the state delegation on its way back stayed for a while in the Kazak district of Taiynshy, where the local authorities had prepared special yurts and food for them. After being very warmly welcomed, and being treated to two glasses of *qymyz*, the guests met the local settled Kazaks introduced to them by the local governor. Stolypin and Krivoshein praised them and said that they were happy people. Stolypin told the Kazaks that he had been sent by the tsar to become better acquainted with their problems. The Kazaks took the opportunity and asked Stolypin to give them the land they had acquired in accordance with sedentary norms as their private property, as well as to enable them to use the remaining lands. They also complained that after becoming sedentary, they had lost their established relationship with relatives, and that other Kazaks, after having witnessed this situation, refused to change their traditional nomadic way of life. They asked Stolypin not to separate them from their relatives and to preserve their traditional social institutions as well after their transition to sedentarization. Stolypin, however, found those arguments not convincing enough, and as the traditional Kazak dish *quyrdaq* was ready, he hurried to enter the house. After the dinner, the Kazaks repeated their requests: for a moment

---

[163] *Zapiska Predsedatelia Soveta Ministrov i Glavnoupravliaiushchego Zemledeliem i Zemleustroistvom*, pp. 80-95.

Stolypin became thoughtful, but gave no answer.[164] Among other things, accompanied by two companies of soldiers and 200 Cossacks specially sent from Omsk, and numerous people who were to look after roads, bridges and carts, as well as hang flags, the state delegation's mission in fact, had been a heavy burden to the local populations, as pointed out by the deputy of the Third Duma, Dziubinskii.[165]

Prompted by Stolypin's emigration campaign, the phenomenon of colonialism became for the first time a subject of closer consideration chiefly due to the publications of the Resettlement Office, and among them especially the journal *Voprosy kolonizatsii* (Questions on Colonization). A wide range of problems discussed by its authors included possible relationships between the metropolis and colonies, the character and principles of colonial policy, the preferable types of colonists, as well as those social and economic infrastructures that were to be fostered in the colonies. Another group of topics related to the manner in which traditional social institutions were to be treated and the degree to which they were to be integrated into central institutions.

Much in the spirit of the time, G. Gins, one of the journal's authors, considered colonization an "unavoidable economic phenomenon which from the point of view of the advanced peoples is a profitable and useful enterprise," and asserted that Russia's civilizing mission in its colonies was to consist of a "policy of a comprehensive cultural development of the unsettled or poorly settled spaces" of the empire that was to imply both the hitherto "poorly used economic resources" and the native inhabitants. Thereby regarded as a "natural" extension of state territories, the geographical location of Russia's colonies was also to lend strong support to maintaining in the colonies the same objectives and social institutions as in the metropolis. Taking France

---

[164] Berdiqojauly, Iskhaq : Jerimiz turasynda körgen bilgenime qosqan pikirlerim, in: *Aiqap,* 6 (1912), pp. 128-131.

[165] Gos. Duma. Stenogr. otchety, sozyv 3, sessiia 4, chast' 3, SPb., 1911, col. 326-336.

as an example, Gins advocated a humane state policy toward the colonized that in respect to the nomads was to facilitate their transition to sedentarization by such measures as the allotment of permanent sedentary plots, financial compensation for their destroyed winter dwellings, etc.[166]

Yet, transformed into the vocabulary of Russian domestic policy, such a human approach, as we have seen, provided little help. Dominated by right-wing Russian nationalists and the Orthodox clergy, the majority of the deputies of the Third Duma (1907-1912) advanced the slogan *zaselenie vazhnee pereseleniia* (resettlement is more important than immigration), supporting thereby Stolypin's policy of land withdrawal and the eviction of nomads from their lands.

Pavel Miliukov, the leader of the Russian Constitutional Democratic party, for his part, characterized the extension to Turkestan of the amendment to Article 120 of the 1891 Steppe Statute legitimizing this policy as "the most cynical law of all laws dictated to the Third Duma by the reactionary government." Nevertheless, sharing the view of the majority of the deputies about the importance of emigration to Russia's domestic policy, he tried to convince the rest of "the peculiar colonizing ability" of Russians who

> in contrast to the English did not build insuperable barriers between themselves and the conquered, but manage to establish a practical and simple relationship with the latter - the ability that has been greeted by our friends in Europe and feared by our enemies. The latter are afraid not of a handful of Russians who will conquer and terrorize but rather of the fact that a local population will become fully imbued with Russians, forming in that way a real, powerful, strategic and political basis for Russia.

As to the fate of the nomads, both Miliukov and the Octobrist, I. I. Kapnist, acknowledged that the so-called nomadic norms represented nothing more than a fiction

---

[166] Gins, G.: Pereselenie i kolonizatsiia, vyp. 1, in: *Voprosy kolonizatsii*, 12 (1913), pp. 2-4, 32.

used by the Resettlement Office to legitimize land withdrawal, and such norms, as far as they concerned the nomadic economy, were impossible to calculate in principle. Only in case of the transition of the nomads to sedentarization could they be allotted land in accordance with the norms used for the Russian peasant settlers. However, this long-term perspective would postpone Russian colonization for an uncertain time. Following Pahlen's footsteps, Miliukov suggested that the "Siberian dregs," who up to recently had represented the main Russian colonizing force in Central Asia, were to be replaced by the "best and healthy representatives of the Russian people," whereas the nomads were to be regarded as a necessary element of the region's geographical environment, and hence their economy was to be supported by the government.[167] All in all, a draft bill introduced by sixty deputies, including the Kadets, the Muslims, the Trudoviki and the Progressists envisaged the organization of land commissions in the seven steppe provinces (including the Zakaspiiskaia oblast') for the definition of land norms for their sedentary, semi-nomadic and nomadic populations that, however, was not approved by the GUZiZ. The speech by its chairman, Glinka, was not marked by originality either:

> It is simply funny to talk about expropriation in regard to the land use of the Kazaks, which implies the violation of property rights. However, since the Kazaks do not have this right, because their lands have been declared state property, they can use them only for the immediate needs of their everyday life. All the remaining lands must be used for resettlement. The amounts of lands allotted to them exceed even those allotted to peasants from the Russian guberniias. We cannot leave them their lands without Russians [settlers] in order to promote their nomadic way

---

[167] Gos. Duma. Stenogr. otchety, sozyv 3, sessiia 4, chast' 1, p. 2607; sessiia 3, chast' 3, col. 3003, SPb., 1910.

of life, but must take decisive measures for the legal resettlement of their lands.[168]
Yet the important arguments cited to the Third Russian Duma by the Muslim deputy, Sadreddin Maksudov, in his two speeches on the issue seem not to have evoked any resonance from other deputies. Pointing out that the unlawful character of the above-mentioned Article 210 of the 1868 Provisional Statute with its declaration of Kazak lands as state property contradicted both Roman and Russian laws, he proved that what was called colonial legislation was in reality a substitution for arbitrary administrative rule: "There is need of a law not because there are excessive lands, but because without the law it would be impossible to take away lands."[169]

Conclusion

The 1916 uprising that embraced all Central Asian regions represented, among other things, a spontaneous response of the indigenous populations to the tsarist land policy. The bitterness with which the rebels had resisted the tsarist military forces and attacked the Russian settles, on the one hand, and the extreme brutality of the punitive forces in destroying peaceful settlements and killing not only the male population but also women and children, on the other, were the most striking features of the uprising.[170] The nomadic Central Asian populations were to provide the majority of recruits, because it was not engaged in cotton growing – the main import product of Russians in Central Asia. Therefore, the strongest resistance and the cruelest

---

[168] Gos. Duma. Stenogr. otchety, sozyv 3, sessiia 2, chast' 3, SPb., 1909, col. 537-538.

[169] Ibid., sozyv 3, sessiia 3, chast' 3, SPb, 1910, col. 3054-3058; see also sozyv 3, sessiia 2, chast' 3, SPb., 1909, col. 569-570.

[170] *Qakharly 1916 jyl. (Qujattar men materialdar jinagy). Groznyi 1916 god. (Sbornik dokumentov i materialov)*, ed. M. Q. Qozybaev, Almaty, 1998, vol. 1, pp. 152-162.

suppression occurred in the nomadic areas.[171] In accordance with the secret report of Kuropatkin to the tsar of February 22, 1917, 2,325 Turkestani Russians were killed, while 1,384 registered as missing.[172] There are, however, no available statistics relating to the general losses of Central Asian indigenous populations. Sanjar Aspendiiarov wrote about "thousands" of indigenes who had perished during the revolt. He also gave a figure of 300,000 Kazaks who fled to China.[173] According to Turar Rysqulov, the Semirech'e natives lost 273,222 or 20.61 percent of their population and one third of their livestock as well.[174] Mustafa Shoqai cited an official Soviet figure of 1,144,000 Central Asian natives who died as a result of famine, the loss of private property and livestock caused by the revolt.[175] In 1917, the military field courts sentenced 347 native participants to death; another 129 were to be imprisoned, while 396 natives were condemned to hard and convict labor.[176]

For the suppression of the revolt in Turkestan, the military forces used 14 battalions, 33 Cossack squadrons, 42 heavy and 69 machine guns. Kuropatkin also reported more than 9,000 households, bridges, roads, churches and hospitals that were destroyed by rebels.[177] Russian soldiers and settlers, in turn, plundered their private property and villages.

The suppression was accompanied by the further seizure of land, which was regarded as punishment. In accordance with Kuropatkin's suggestion made at a meeting of October 16, 1916, "all lands where Russian blood has been shed are to be taken away from the

---

[171] Ryskulov, T.: *Vosstanie tuzemtsev v Srednei Azii v 1916 godu*, Kyzyl Orda, 1927, p. 9.
[172] *Qakharly 1916 jyl*, vol. 1, p. 336.
[173] Aspendiiarov, S.: *Qazaq tarikhynyn ocherkteri*, Almaty, 1994, pp. 108-109.
[174] Ryskulov, T.: *Vosstanie tuzemtsev v Srednei Azii v 1916 godu*, pp. 61-62.
[175] Chokaev, M.: *Turkestan pod vlast'iu Sovetov (K kharakteristike diktatury proletariata)*, Parizh, 1935, p. 78.
[176] Aspendiiarov, S.: *Qazaq tarikhynyn ocherkteri*, p. 97.
[177] *Qakharly 1916 jyl*, vol. 1, p. 336.

Kirgizs." Consequently, 2,510,361 desiatinas of land in Turkestan were withdrawn for the establishment of new 5 Cossack *stanitsas*.[178]

By 1917 forty five million hectares of Kazak pastures had been withdrawn by the tsarist government, which made up sixteen percent of all Kazak lands, while the number of Slavic settlers resettled on them reached 23 percent of the whole steppe population.[179]

After 1917 land and land policy remained crucial for the integration of the nomadic Kazaks into the Soviet empire. For the tsarist government the sedentarization of the Kazaks had been thinkable only as their transformation into Russian *muzhiks*. As for the Soviets, they were able to view sedentarization only as a transformation of Kazak herders into Soviet *kolkhozniki*. By contrast to the tsarist government, the Bolsheviks added an ideological character to the campaign for sedentarization; the mass collectivization of agriculture in the 1930s was connected with the final victory of Socialism in the Soviet East. The Kazaks finally became sedentary in the bounds of their own republic, but it cost them the loss of 42 percent of their population, hundreds of thousands of whom migrated while millions died from violence and starvation. The loss of 80 percent of their livestock and in the end the complete destruction of their traditional economy and social structures were also the outcome of Soviet-style sedentarization.

Collectivization as well as the following campaigns of industrialization and the 1950s development of Virgin Lands were accompanied by the intensified mass immigration of European and non-European populations (deported peoples, workers, peasants, etc.). Kazakhstan has experienced a powerful pressure of peoples' immigration unlike any other among the former Soviet republics. In 1979 the Kazaks made up only 33 percent of the republic's population. New "colonial manipulations of space" (*natsional'noe raionirovanie* – literally, national districtization) included further with-

---

[178] Ibid., p. 85.
[179] Qoigeldiev, M.: *Alash qozgalsy*, p. 50.

drawal of land for building enterprises, *kolkhozy* (collective farms), but also for military use, including the testing of atomic weapons.

Bibliography

Abdirov, M. J.: *Istoriia kazachestva Kazakhstana*, Almaty, 1994

Aiqap (newspaper), 6 (1912)

Aqqulyuly, S. (ed.): Bökeikhan, Alikhan: Tandamaly (Izbrannoe): *Nauchnye issledovaniia, trudy na kazakhskom i russkom iazykakh: monografii, stat'i, rechi, doklady, stikhi, perevody, pis'ma*, Almaty, 1995

Aspendiiarov, S.: *Qazaq tarikhynyn ocherkteri*, Almaty, 1994

Bekmakhanov, E. B. *Prisoedinenie Kazakhstana k Rossii*, M., 1957

Bekmakhanova, N. E.: *Mnogonatsional'noe naselenie Kazakhstana i Kirgizii v epokhu kapitalizma (60-gody XIX v. – 1917 g.)*, M., 1986

Beskrovnyi, L. G.: *Russkaia armiia i flot v XIX veke*, M., 1973

Borodin, A.: *Ural'skoe kazach'e voisko. Statisticheskoe obozrenie v 2-kh tomakh*, vol. 1, Ural'sk, 1891

Brower, Daniel R./ Lazzerini, Edward (eds.): *Russia's Orient: Imperial Borderlands and Peoples, 1700-1917*, Bloomington/Indianapolis: Indiana University Press, 1997

The Central State Archive of the Republic of Kazakstan (TsGA RK), op. 1, d. 18, l. 9

Chokaev, Mustafa: *Turkestan pod vlast'iu Sovetov (K kharakteristike diktatury proletariata)*, Parizh, 1935

Diakin, V. S.: *Natsional'nyi vopros vo vnutrennei politike tsarizma (XIX – nachalo XX vv.)*, SPb., 1998

Eleuov, T. E. (ed.): *Voprosy istorii Kazakhstana XIX – nachala XX veka*, Alma-Ata, 1961

Glebov, A.: *Chto mogut dat' pereseleniia krest'ianstvu*, SPb., 1907

Gosudarstvennaia Duma. Stenograficheskie otchety, sozyv 3, sessiia 2, chast' 3, SPb., 1909; sozyv 3, sessiia 4, chast' 1; sozyv 3, sessiia 3, chast'3, SPb., 1910; sozyv 3, sessiia 4, chast' 3, SPb., 1911;

Iskhakov, S. M.: *Pervaia mirovaia voina glazami rossiiskikh musul'man* (unpublished manuscript)

Istoriia Kazakhskoi SSR, vol. 3, Alma-Ata, 1979

Jangalin, M. O./Kireev, F. N./Shakhmatov, V. F. (eds.): *Kazakhsko-russkie otnosheniia v XVIII – XIX vekakh (1771-1867 gody) (Sbornik dokumentov i materialov)*, Alma-Ata, 1964

Kastelianskii, A. I. (ed.): *Formy natsional'nogo dvizheniia v sovremennykh gosudarstvakh. Avstro-Vengriia. Rossiia. Germaniia*, SPb., 1910

Katanaev, G. E.: *Kirgizskii vopros v Sibirskom kazach'em voiske*, Omsk, 1904

Kaufman, A. A.: *K voprosu o russkoi kolonizatsii Turkestanskogo kraia. Otchet chlena Uchenogo komiteta Kaufmana po komandirovke letom 1903 g.*, SPb., 1903

Kaufman, A. A.: *Materialy po voprosu ob organizatsii rabot po obrazovaniiu pereselencheskikh uchastkov v stepnykh oblastiakh (Iz otcheta starshego proizvoditelia rabot Kaufmana po komandirovke v Akmolinskuiu oblast' letom 1897 g.)*, SPb., 1897

Los'-Roshkovskii, F.: Khodokam i pereselentsam, napravliaiushchimsia v Kustanaiskii uezd Turgaiskoi oblasti (Nastavleniia, opisaniia raiona i uchastkov), Poltava, 1912

Masevich, M. G. (ed.): *Materialy po istorii politicheskogo stroia Kazakhstana (so vremeni prisoedineniia Kazakhstana k Rossii do Velikoi Oktiabr'skoi Sotsialisticheskoi revoliutsii)*, Alma-Ata, 1960

Marks, Steven, G.: *Road to Power. The Trans-Siberian Railroad and the Colonization of Asian Russia, 1850-1917*, Ithaca/New York: Cornell University Press, 1991

Materialy po istorii Kazakhskoi SSR (1785-1828 gg.), vol. IV, M./L., 1940

Materialy po kirgizskomu zemlepol'zovaniiu, sobrannye i razrabotannye ekspeditsiiei po issledovaniiu stepnykh oblastei, vol. I: Akmolinskaia oblast', Kokchetavskii uezd, Voronezh, 1898; vol. IV: Semipalatinskaia oblast', Pavlodarskii uezd, Voronezh, 1903; vol. VII: Turgaiskaia oblast'. Aktiubinskii uezd, Voronezh, 1903; vol. VIII: Semipalatinskaia oblast', Zaisanskii uezd, SPb., 1909; vol. XII: Akmolinskaia oblast', Petropavlovskii uezd, Chernigov, 1908

Orazaev, F. M. (ed.): *Tsarskaia kolonizatsiia v Kazakhstane (Po materialam russkoi periodicheskoi pechati XIX veka)*, Almaty, 1995

Otchetnye dannye po Akmolinskomu pereselencheskomu raionu za 1907 g. na osnove otcheta byvshego zav. Pereselencheskim delom v Akmolinskom raione Reznichenko, vyp. 49, SPb., 1908

Pahlen, Constantin Graf von der Pahlen: *Im Autrag des Zaren in Turkestan 1908-1909*, Stuttgart, 1969 [Bibliothek klassischer Reiseberichte]

Pahlen. K. K.: *Pereselencheskoe delo v Turkestane. Otchet po revizii Turkestanskogo kraia, proizvedennyi po vysochaishemu poveleniiu senatorom gofmeisterom grafom K. K. Palenom*, SPb., 1910

Pamiatnaia knizhka Semipalatinskoi oblasti na 1901 god, vyp. V, Semipalatinsk, 1901

Pereselenie v Stepnoi krai v 1906 g. (oblasti Akmolinskaia i Semipalatinskaia), vyp. 27

Qakharly 1916 jyl (Qujattar men materialdar jinagy). Groznyi 1916 god (Sbornik dokumentov i materialov), ed. M. Q. Qozybaev, vol. 1, Almaty, 1998

Qoigeldiev, Mambet: *Alash qozgalysy*, Almaty, 1995

Rech' (newspaper), 1906, N 112

Rumiantsev, P.: *Materialy po obsledovaniiu tuzemnogo i russkogo starozhil'cheskogo khoziaistva i zemledeliia v Semirechenskoi oblasti*, vol. 3, SPb., 1912

Ryskulov, T.: *Vosstanie tuzemtev v Srednei Azii v 1916 godu*, Kyzyl Orda, 1927

The Russian State Historical Archive (RGIA), f. 391, op. 4, d. 1663; f. 1276, op. 4, d. 468, l. 300, 211-212, 218

Sbornik zakonov i rasporiazhenii po pereselencheskomu delu i pozemel'nomu ustroistvu v guberniiakh i oblastiakh Aziatskoi Rossii (po 1 avgusta 1909 g.), SPb., 1909

Sedel'nikov, T. (deputat I Gosdumy): *Bor'ba za zemliu v kirgizskoi stepi (Kirgizskii vopros i konizatsionnaia politika pravitel'stva)*, SPb., 1907

Shonauly, Tel'jan: *Jer tagdyry - el tagdyry*, Almaty, 1995

Suleimenov, B. S.: *Revoliutsionnoe dvizheniie v Kazakhstane v 1905-1907 godakh*, Alma-Ata, 1977

Suleimenov, B. S./Basin, V. Ia.: *Kazakhstan v sostave Rossii v XVIII - nachale XX veka*, Alma-Ata, 1981

Tregubov, A. L.: *Pereselencheskoe delo v Semipalatinskoi i Semirechenskoi oblastiakh. Vpechatleniia i zametki chlena Gos. Dumy A. L. Tregubova po poezdke letom 1909 g.*, SPb., 1910

Tresviatskii, V. A. (ed.): *Materialy po zemel'nomu voprosu v Aziatskoi Rossii, vyp. 1: Stepnoi krai*, Petrograd, 1917; vyp. 6 : *Itogi pereselencheskogo dela za Uralom s 1906 po 1915 gg.*, Petrograd, 1918

V mire musul'manstva (newspaper), 1911, no. 27

Voprosy kolonizatsii (journal), 12 (1913)

Zapiska Predsedatelia Soveta Ministrov i Glavnoupravliaiushchego Zemledeliem i Zemleustroistvom o poezdke v Sibir' i Povolzh'e v 1910 godu. Prilozhenie k vsepoddanneishemu dokladu, SPb., 1910

Zhurnal Soveshchaniia o poriadke kolonizatsii Semirechenskoi oblasti, Vernyi, 1908

Bei Fragen zur Produktsicherheit wenden Sie sich bitte an:
If you have any questions regarding product safety,
please contact:

Walter de Gruyter GmbH
Genthiner Straße 13
10785 Berlin
productsafety@degruyterbrill.com